THE MEDITERRANEAN

Other Titles in
ABC-CLIO'S
NATURE AND HUMAN SOCIETIES SERIES

FORTHCOMING

Australia, New Zealand, and the Pacific, Donald S. Garden

Northeast and Midwest United States, John T. Cumbler

Northern Europe, Tamara L. Whited, Jens I. Engels,
Richard C. Hoffmann, Hilde Ibsen, and Wybren Verstegen

Sub-Saharan Africa, Gregory H. Maddox

NATURE AND HUMAN SOCIETIES

THE MEDITERRANEAN
An Environmental History

J. Donald Hughes

Santa Barbara, California   *   Denver, Colorado   *   Oxford, England

Library of Congress Cataloging-in-Publication Data
Hughes, J. Donald (Johnson Donald), 1932–
    The Mediterranean : an environmental history / J. Donald Hughes.
        p.   cm. — (Nature and human societies)
    Includes bibliographical references and index.
    ISBN 1-57607-810-8 (hardback : alk. paper) — ISBN 1-57607-811-6 (eBook)
1. Human ecology—Mediterranean Region.   2. Mediterranean Region—
Environmental conditions.   I. Title.   II. Series.
    GF541.H833 2005
    304.2'0182'2—dc22

                                                            2004029969

06 05 04 03   10 9 8 7 6 5 4 3 2 1

This book is also available on the World Wide Web as an eBook. Visit abc-clio.com for details.

ABC-CLIO, Inc.
130 Cremona Drive, P.O. Box 1911
Santa Barbara, California 93116-1911

This book is printed on acid-free paper ∞ .
Manufactured in the United States of America

CONTENTS

SERIES FOREWORD

Long ago, only time and the elements shaped the face of the earth, the black abysses of the oceans, and the winds and blue welkin of heaven. As continents floated on the mantle they collided and threw up mountains, or drifted apart and made seas. Volcanoes built mountains out of fiery material from deep within the earth. Mountains and rivers of ice ground and gorged. Winds and waters sculpted and razed. Erosion buffered and salted the seas. The concert of living things created and balanced the gases of the air and moderated Earth's temperature.

The world is very different now. From the moment our ancestors emerged from the southern forests and grasslands to follow the melting glaciers or to cross the seas, all has changed. Today the universal force transforming the earth, the seas, and the air is for the first time a single form of life: we humans. We shape the world, sometimes for our purposes and often by accident. Where forests once towered, fertile fields or barren deserts or crowded cities now lie. Where the sun once warmed the heather, forests now shade the land. One creature we exterminate, only to bring another from across the globe to take its place. We pull down mountains and excavate craters and caverns, drain swamps and make lakes, divert, straighten, and stop rivers. From the highest winds to the deepest currents, the world teems with chemical concoctions only we can brew. Even the very climate warms from our activity.

And as we work our will upon the land, as we grasp the things around us to fashion into instruments of our survival, our social relations, and our creativity, we find in turn our lives and even our individual and collective destinies shaped and given direction by natural forces, some controlled, some uncontrolled, and some unleashed. What is more, uniquely among the creatures, we come to love the places we live in and know. For us, the world has always abounded with unseen life and manifest meaning. Invisible beings have hidden in springs, in mountains, in groves, in the quiet sky and in the thunder of the clouds, and in the deep waters. Places of beauty from magnificent mountains to small, winding brooks have captured our imaginations and our affection. We have perceived a mind like our own, but greater, designing, creating, and guiding the universe around us.

The authors of the books in this series endeavor to tell the remarkable epic of the intertwined fates of humanity and the natural world. It is a story only now coming to be fully known. Although traditional historians told the drama of men and women of the past, for more than three decades now many have added the natural world as a third actor. Environmental history by that name emerged in the 1970s in the United States. Historians quickly took an interest and created a professional society, the American Society for Environmental History, and a professional journal, now called *Environmental History.* American environmental history flourished and attracted foreign scholars. By 1990 the international dimensions of the field were clearly growing; European scholars joined together to create the European Society for Environmental History in 2001, with its journal, *Environment and History.* Then, in 2004, scholars meeting in Havana, Cuba, organized La Sociedad Latinoamericana y Caribeña de Historia Ambiental. With an abundant and growing literature of world environmental history now available, a true world environmental history can appear.

This series is organized geographically into regions determined as much as possible by environmental and ecological factors, and secondarily by historical and historiographical boundaries. Befitting the vast environmental historical literature on the United States, four volumes tell the stories of the North, the South, the Plains and Mountain West, and the Pacific Coast. Other volumes trace the environmental histories of Canada and Alaska, Latin America and the Caribbean, Northern Europe, the Mediterranean region, sub-Saharan Africa, Russia and the former Soviet Union, South Asia, Southeast Asia, East Asia, and Australia and Oceania. Authors from around the globe, experts in the various regions, have written the volumes, almost all of which are the first to convey the complete environmental history of their subjects. Each author has, as much as possible, written the twin stories of the human influence on the land and of the land's manifold influence on its human occupants. Every volume contains a narrative analysis of a region along with a body of reference material. This series constitutes the most complete environmental history of the globe ever assembled, chronicling the astonishing tragedies and triumphs of the human transformation of the earth.

Creating the series, recruiting the authors from around the world, and editing their manuscripts has been an immensely rewarding experience for me. I cannot thank the authors enough for all of their effort in realizing these volumes. I owe a great debt too to my editors at ABC-CLIO: Kevin Downing, who first approached me about the series and helped me get it going; and Steven Danver, who has shepherded the volumes through delays and crises to publication. Their unfaltering support for and belief in the series were essential to its successful completion.

—Mark Stoll
Department of History
Texas Tech University
Lubbock, Texas

PREFACE AND ACKNOWLEDGMENTS

I have been a student of the Mediterranean for most of my life. Born in Santa Monica, in the Mediterranean-like coastal climate of Southern California, I came to regard a seasonal regime very much like that of the Mediterranean as proper, and vegetation such as the chaparral, the Californian equivalent of maquis, as normal. I studied botany there, learning about many Mediterranean plants, and received a degree in biology, with an emphasis on botanical genetics, at the University of California at Los Angeles. While I was working on my PhD in history at Boston University, I made my first visit to the Mediterranean area in 1959, and I have returned many times since, visiting the lands from Portugal in the west to Egypt and Jordan in the east. During a year in residence at the American School of Classical Studies in Athens, Greece, in 1966–1967, I began my study of the environmental history of ancient Greece, Rome, Egypt, and Mesopotamia, a study that produced a book, *Ecology in Ancient Civilizations*, published in 1975 (Albuquerque: University of New Mexico Press). Continued involvement in that scholarly field eventually resulted in what one might well call a sequel, *Pan's Travail: Environmental Problems of the Ancient Greeks and Romans*, in 1994 (Baltimore: Johns Hopkins University Press). Meanwhile, I wrote a number of articles and chapters on related subjects, among which I would mention the practice of setting aside and preserving sacred groves, where I found interesting parallels between the ancient Mediterranean and India, both ancient and modern. I also included sections on the Mediterranean in my most recent book, *An Environmental History of the World: Humankind's Changing Role in the Community of Life*, published in 2001 (London and New York: Routledge).

The material and conclusions in this book were made possible by research assistance from a number of sources. Among these I must mention and thank the University of Denver, including the Department of History and its chairpersons, John Livingston, Michael Gibbs, and Ari Kelman; the Division of Arts, Humanities, and Social Sciences, with its deans, Roscoe Hill and Gregg Kvistad, for liberal support of many kinds; and Chancellor Dan Ritchie and Provosts

Wiliam Zaranka and Robert D. Coombe for support given through the John Evans Professorship. Thanks also for a grant and for continuing encouragement from the Charles A. and Anne Morrow Lindbergh Foundation and its president, Reeve Lindbergh. Alberto Vieira and O Centro de Estudos de História do Atlântico in Madeira also deserve gratitude for their aid, as do Marcus Hall and the European University Institute in Florence, and Niki Goulandris and the Goulandris Natural History Museum in Kifisia, Greece.

Environmental history is not only a subject; it is also an endeavor and a vocation. Colleagues in my professional associations, the American Society for Environmental History and the European Society for Environmental History, have offered advice and scholarly criticism and encouragement. Among these I must express especial indebtedness to Mauro Agnoletti, Karl W. Butzer, John Dargavel, Richard Grove, Lorne Hammond, Richard C. Hoffmann, Elisabeth Johann, John R. McNeill, Marc Pavé, Christian Pfister, Paolo Squatitri, Mark Stoll, Petra J. E. M. Van Dam, Paolo Visona, Douglas Weiner, Verena Winiwarter, and Donald Worster. I am also greatly indebted to James O'Connor and Barbara Laurence, who through their journal, *Capitalism, Nature, Socialism*, gave me the opportunity to try out many of my ideas in a regular column, "Ripples in Clio's Pond."

Finally, many thanks to my wife, Pamela Hughes, who urged me to accept the invitation to write this book, revived my flagging spirits when it seemed I might never complete it, protected my time so I could devote the necessary hours to it, and listened when I needed to talk about it.

INTRODUCTION

The provinces bordering on the . . . basins of the Mediterranean enjoyed a healthfulness and an equability of climate, a fertility of soil, a variety of vegetable and mineral products, and natural facilities for the transportation and distribution of exchangeable commodities, which have not been possessed in an equal degree by any territory of like extent in the Old World or the New.
—GEORGE PERKINS MARSH, *MAN AND NATURE*, 1864

The Mediterranean landscape bears the scars of ancient wounds.
—JOHN R. MCNEILL, *THE MOUNTAINS OF THE MEDITERRANEAN WORLD*, 1992

In writing a book about the environmental history of the Mediterranean region, it seems only fair to the reader to begin with a definition of environmental history, and an explanation of what I intend to include in the geographical area of the Mediterranean. This is true because after some thirty years of writing about the subject, I still find that "What is environmental history?" is the prevailing question I get whenever I tell anyone outside the historical profession that I am an environmental historian, in spite of the fact that environmental history has had an organized society and a journal for more than a quarter of a century. I also find that no two history books on the Mediterranean region agree completely on where the boundaries of that region are located.

The first question can be answered in more than one way, but here is what I would say: *As a subject,* environmental history is the study of the interaction between human societies and the natural environment through time. *As a method,* it is the use of ecological analysis as a means of understanding human history. Environmental historians recognize the ways in which the living and nonliving systems of the Earth have influenced the course of human affairs. They also evaluate the impacts of changes caused by human agency in the natural environment. Among the kinds of human activities it looks at are those that provide basic sustenance, such as hunting, gathering, fishing, herding, and agriculture. Others deal with the organization of human settlements from villages to cities, including the provision of basic materials by water management, forestry, mining, and metallurgy. Technology and industries, affecting most human activities including warfare, have become more sophisticated and taken up more human energy as the centuries have passed. All of these affect the natural environment in many ways—both positively and negatively, from human points of view. Many of them make the environment more amenable to human use. But all cause other changes that can be damaging, such as deforestation, reduction of biodiversity through extinctions, desertification, salinization, and pollution. In recent decades, newly recognized damaging changes include radioactive fallout, acid precipitation, and global warming due to the greenhouse effect. These in turn make the environment less amenable to sustained human use. Societies often have tried to accent the positive changes and limit the negative ones through pollution control and conservation of natural resources, including the preservation of certain designated areas and the protection of endangered species.

Another aspect of environmental history is the study of human thought about the natural environment and attitudes toward it, including the study of nature, the science of ecology, and the ways in which systems of thought such as religions, philosophies, political ideologies, and popular culture have affected human treatment of various aspects of nature. It is impossible to understand what has happened to the Earth and its living systems without giving attention to this aspect of social and intellectual history. Environmental history must be perceptive of human interconnections in the world community, and of the interdependence of humans and other living beings on the planet. Environmental history takes an interdisciplinary approach that uses traditional economic, social, and political forms of historical analysis, but it relates them to the insights of sciences such as ecology and geography. The environment can no longer be seen as the stage setting on which human history is enacted. It is an actor; indeed, it comprises a major portion of the cast.

What needs emphasis is that all human societies, everywhere, throughout history, have existed within and depended upon biotic communities. This is true of huge cities as well as small farming villages and hunter clans. The connectedness of life is a fact. Humans never existed in isolation from the rest of life, and could not exist alone, because they are only one part of the complex and intimate associations that make life possible. The task of environmental history is the study of human relationships, through time and subject to frequent and often unexpected changes, with the natural communities of which they are part.

A typical Mediterranean scene with sea and land in intimate connection. The photograph shows the bay of Epidaurus, Greece, with the peninsula of Methone in the background. Geraniums in the foreground, a native Mediterranean species, take their name from geranos, *Greek for "crane," referring to the long beaklike point on the fruits. (Photo courtesy of J. Donald Hughes)*

The idea of the environment as something separate from the human, and offering merely a setting for human history, is misleading. The living connections of humans to the communities of which they are part must be integral components of the historical account. Whatever humans have done to the rest of the community has inevitably affected themselves. To a very large extent, ecosystems have influenced the patterns of human events. We, in turn, have to an impressive degree made them what they are today. That is, humans and the rest of the community of life have been engaged in a process of coevolution that did not end with the origin of the human species, but has continued to the present day. Historical writing should not ignore the importance and complexity of that process.

And now to determine what should be included in the Mediterranean region. First of all, it is important to note that the Mediterranean is a *geographical area centered on a sea*, not a continent. Indeed, the lands contiguous to the Mediterranean Sea include parts of the three great continents: Europe, Africa, and Asia. But just how much sea should be included in our definition? The Black Sea is connected to the Mediterranean, and its waters flow into the larger sea through the Bosporus and the Dardanelles. It seems right to include it, but its climate and culture, especially in the north, and its water budget are distinct. The watershed of the Mediterranean and Black seas can hardly be a useful boundary, however, since that would of necessity include much of the Sudan, Ethiopia, Uganda, Rwanda, and Burundi by way of the Nile; too much of France by the Rhône; and vast areas of central Europe, Ukraine, and Russia by the Danube, Dnieper, and Don. At the other extreme, it would exclude a very large proportion of Spain and all of Portugal, whose rivers flow into the Atlantic Ocean, and a lot of North Africa, since much of its dry coast has no rivers at all.

The Mediterranean *climatic zone* seems to offer a good limitation unless one tries to define it too closely. Climatic boundaries lack sharpness. Do rainy Dalmatia and bone-dry Libya really share the same climate? It has often been remarked that the area in which olives can be grown defines the Mediterranean climatic zone. It is true that the places where olives grow are the most typically

An old olive tree, damaged by fire, puts out new shoots at the archaeological site of Oppido Mamertina in Calabria, Italy. Some geographers hold that the distribution of this domesticated tree defines the limits of the typical Mediterranean climate. (Photo courtesy of J. Donald Hughes)

"Mediterranean" by almost every standard and can be recognized as parts of the Mediterranean heartland. The olive, after all, has long been regarded as a sacred plant by the Mediterranean peoples. The victors in the Olympic Games received olive crowns, the kings of Israel were designated by anointing with olive oil, and the oldest Muslim university in Tunisia is named al-Zitouna, "The Olive Tree." Some districts such as the Libyan coast are too arid for the olive tree, however, and other districts are too cool, such as many highlands, northern Spain and Italy, and the interior of Turkey. These are districts that demand to be included in the Mediterranean region for many reasons.

Political and cultural definitions have their drawbacks as well. One could define the Mediterranean world as containing all those nations that have Mediterranean seacoasts, but that would exclude Portugal, which faces the Atlantic but is very Mediterranean in character, and it would include all of France, but most of France is not at all Mediterranean in climate or culture. Also, Libya and Algeria include huge chunks of the Sahara, and only the northernmost margin of the Sahara can be admitted within the Mediterranean realm.

In defining the Mediterranean region for this book, therefore, I have to be somewhat arbitrary, and I have decided to leave the boundary permeable or even hazy, while knowing that my choices are open to criticism. Some areas can be included for one purpose and excluded for another. In Europe, therefore, I include Portugal, Spain, the southern part of France, all of Italy, the coastal districts of the former Yugoslavia, Albania, all of Greece, and the European part of Turkey. In Africa, the northern coastal plain and the mountainous part of the Maghreb in Morocco, Algeria, and almost all of Tunisia can be regarded as Mediterranean, along with coastal Libya, and Egypt with the exception of Egypt's Saharan interior. In Asia, Turkey, Syria, Lebanon, Jordan, Palestine, and Israel definitely merit inclusion. I also include Iraq and Kuwait for historical and cultural reasons, although the Tigris and Euphrates rivers flow to the Persian Gulf, and the Mesopotamian climate is marginally Mediterranean, if at all. To return to the Mediterranean Sea itself, I would of course include all the islands that its waters enclose: Cyprus, Crete, and the myriad smaller Greek Islands; the islands of the Adriatic coast; Malta; Sicily; the Aeolian Islands; Sardinia; Corsica; the Balearics; and a number of smaller islands here and there. In addition, I reach out to the Atlantic to annex for my purposes the islands of Macaronesia: the Canaries and Madeira, because they have a Mediterranean climate, and in addition to having been colonized by Mediterranean nations, they illustrate important aspects of the environmental history of Mediterranean islands.

There are some other districts that, while beyond the Mediterranean proper, have to be included briefly for some purposes. Among these are the nations bordering the Black Sea coastal zones: Bulgaria, Romania, Ukraine, Russia, and

A terraced hillside on the Greek island of Samos permits the growth of grapevines on a steep slope, while limiting the amount of erosion that would otherwise occur. (Photo courtesy of J. Donald Hughes)

Georgia (I have already mentioned and included Turkey, a major Black Sea state). Other nations that influence the story and enter it at times are Saudi Arabia and the Persian Gulf states of Oman, Qatar, Bahrain, Kuwait, United Arab Emirates, and Iran (Persia in earlier times). The case study on Egypt and the Aswan dams must take important account of the nations on the Nile upstream from Egypt: Sudan, Ethiopia, and several other East African states.

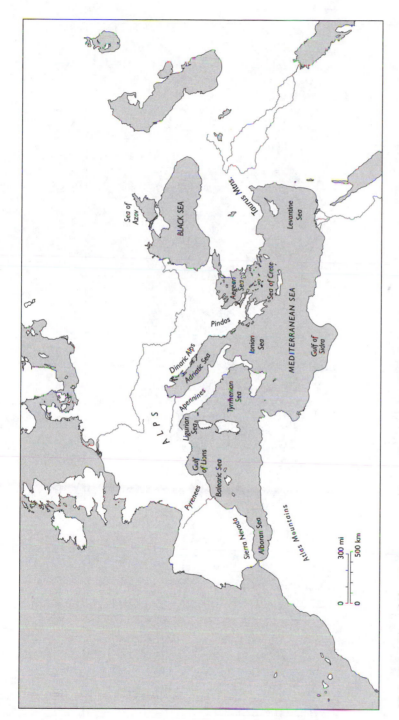

Major Divisions of the Mediterranean Sea and Important Mountain Ranges of the Mediterranean Region.

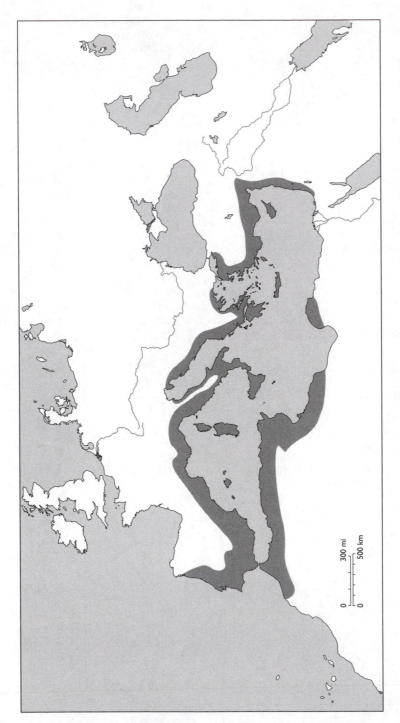

300 mi

500 km

Range of Cultivation of the Olive Tree in the Mediterranean Area from Classical Greek and Roman Times to the Present.

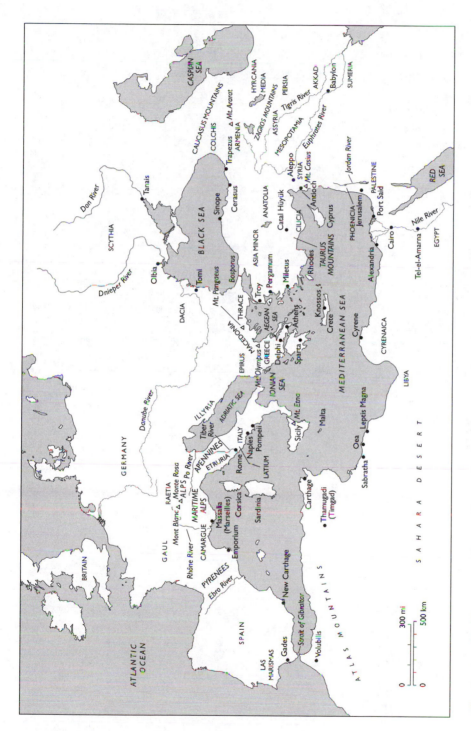

The Ancient Mediterranean during the Time of the Roman Empire, from the First to Fifth Centuries AD.

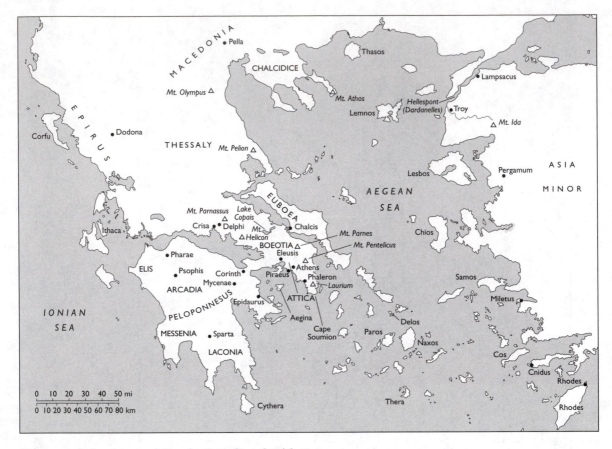

Greece and the Aegean from the Fourth and Fifth Centuries BC.

Italy, Emphasizing Roman and Renaissance Times.

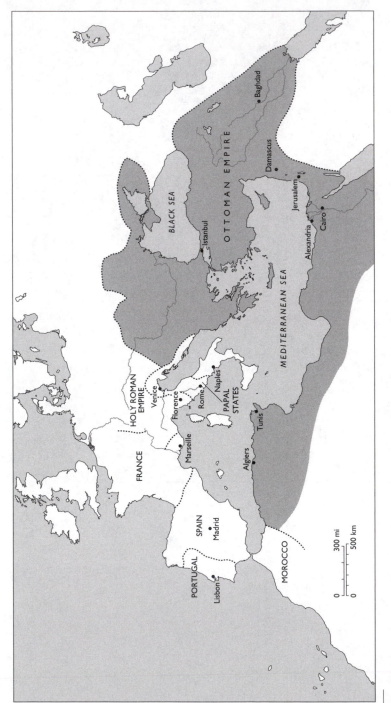

The Ottoman Empire at Its Greatest Extent in the Sixteenth Century.

PORTUGAL

Lisbon

SPAIN

Madrid

MOROCCO

0 300 mi

0 500 km

FRANCE

Marseille

Algiers

Tunis

HOLY ROMAN
EMPIRE

Venice

Florence

Rome

Naples

PAPAL
STATES

MEDITERRANEAN SEA

BLACK SEA

Istanbul

OTTOMAN EMPIRE

Baghdad

Damascus

Jerusalem

Alexandria

Cairo

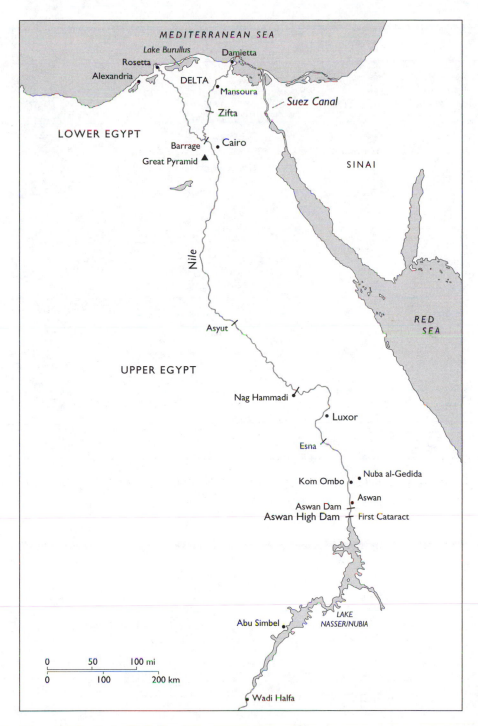

MEDITERRANEAN SEA

Lake Burullus

Damietta

Rosetta

Alexandria

DELTA

Mansoura

SINAI

Suez Canal

Zifta

LOWER EGYPT

Barrage

Cairo

Great Pyramid ▲

Nile

Asyut

UPPER EGYPT

RED
SEA

Nag Hammadi

Luxor

Esna

Kom Ombo

Nuba al-Gedida

Aswan Dam

Aswan

Aswan High Dam

First Cataract

Abu Simbel

*LAKE
NASSER/NUBIA*

0	50	100 mi
0	100	200 km

Wadi Halfa

Egypt in the Twentieth Century.

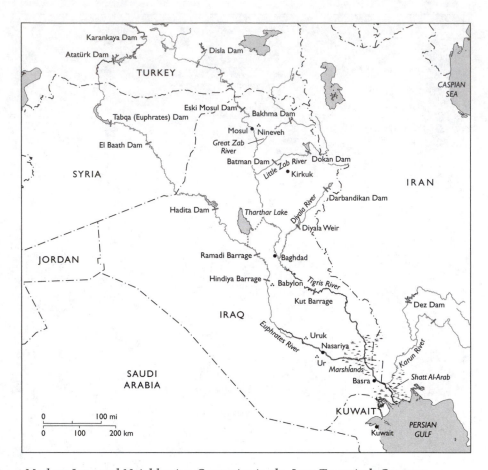

Modern Iraq and Neighboring Countries in the Late Twentieth Century.

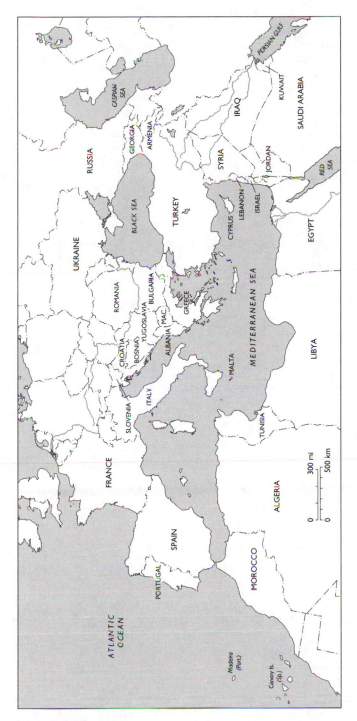

Present-Day Countries of the Mediterranean Area.

THE MEDITERRANEAN ENVIRONMENT AND ITS FIRST HUMAN INHABITANTS

All the world is heir of the Mediterranean. All the world is her debtor.

—ELLEN CHURCHILL SEMPLE, *THE GEOGRAPHY OF THE MEDITERRANEAN REGION*, 1931

Woe betide the historian who thinks that this preliminary interrogation is unnecessary, that the Mediterranean as an entity needs no definition because it has long been clearly defined [and] is instantly recognizable.

—FERNAND BRAUDEL, *THE MEDITERRANEAN AND THE MEDITERRANEAN WORLD*, 1972

The Mediterranean area is defined by its relationship to a sea rather than as a continental region. In the course of a dynamic series of tectonic movements over millions of years, the sea came to be surrounded by the coasts of southern Europe, western Asia, and northern Africa, most of which are mountainous, as a theater's orchestra is surrounded by towering banks of seats. Humans appeared in this theater of life as they moved northward from the sub-Saharan African scene of human origins. They encountered unfamiliar species of animals and quickly became their hunters. They learned the lore of the use of plants, birds, and insects. Hunters and gatherers ranged throughout the Mediterranean region from at least about 200,000 BC onward. They killed animals and gathered plants, exhausting the supply locally in many places, but in a long process of trial and error, cultural traditions evolved that limited their impacts on their natural surroundings and tended toward balance, however precarious. Around 15,000 BC small groups near the Nile River in Egypt and perhaps also along the Tigris and Euphrates rivers in Mesopotamia began to experiment with planting seeds and harvesting crops. Others in the mountains along the margins of the Near East domesticated sheep, goats, and other species of animals. By 6000 BC, this "Agricultural Revolution" had spread through Asia Minor to Greece and other lands around the Eastern Mediterranean, and within another thousand years had reached Italy. Farming villages dotted the most fertile lands, and herders pastured their flocks where they could find grass, leading them up to the highlands in summer and returning to the warmer lowlands in winter, a process called transhumance. The transformation of the Mediterranean landscape by human agency, through the hunting of wild animals, the expansion of agricultural lands at the expense of forest, the use of fire, and the intensive grazing of the herds, was already advancing. Indeed, it took place over such a long period of time that it may be said that the Mediterranean environment and its human inhabitants literally evolved together. For many thousands of years, the Mediterranean has been an inhabited landscape, shaped to a great extent by its inhabitants. The peoples of the Mediterranean have been at least equally shaped by their environments.

THE MEDITERRANEAN SETTING

For the purposes of this volume, the Mediterranean region is defined as the entire Mediterranean Sea, including the Black Sea, and the adjoining lands that have a Mediterranean-type climate and its associated ecosystems, or did within historical times (refer to the Introduction for a complete listing of modern-day lands falling within this region). The sea is the central element of the Mediterranean region and a contributing influence to its climate. The Mediterranean Sea extends 3,600 kilometers (2,300 miles) from Gibraltar to Lebanon. Its north-south dimension is extremely uneven, but the maximum distance, from Venice to Libya, is about 1,100 kilometers (680 miles). In area, it covers 2.5 million square kilometers (970,000 square miles), or if the Black Sea is included, 2.95 million square kilometers (1.14 million square miles). The coastline is lengthy and irregular, so that there is an intimate contact between land and sea, and an extensive littoral zone. Italy's coastline, for example, is 6,660 kilometers (4,140 miles) long; Greece, though smaller than Italy, has a coastline more than twice as long, at 15,000 kilometers (9,300 miles), and there is no place in Greece further than 110 kilometers (68 miles) from the seacoast.

The Mediterranean Sea is almost completely landlocked, a fact that has many important effects on the environment in the region. Its only natural connection to the great oceans of the world is through the Strait of Gibraltar, which separates Europe and Africa by a stretch of water just under 14 kilometers (9 miles) wide at the narrowest point, and only 365 meters (1,200 feet) deep at the deepest. This effectively isolates the Mediterranean from the tides of the world ocean, so that its tides are only those that result from the gravitational forces of the moon and sun on the sea itself—along most shores, less than a meter between high and low. This makes the sea level in ports stable from day to day and makes building and cultivation possible relatively close to the shoreline. The Mediterranean is in a warm, relatively dry region, so the rate of evaporation from its surface is high, averaging 145 centimeters (57 inches) annually. Rainfall replaces 27 percent of this, the inflow of the Nile and all other rivers 6 percent, and another 6 percent comes in from the Black Sea, which receives more water from rivers than it loses through evaporation. The rest, 61 percent, must flow inward through the straits from the Atlantic. These waters are relatively warm, and the Mediterranean is further heated by the sun and made more saline by evapora-

The active volcanism of Mediterranean geology is illustrated by this view of the smoking peak of Mount Etna, seen from the ancient theater of Taormina, Sicily. (Photo courtesy of J. Donald Hughes)

tion. The more saline waters, being denser, sink to the bottom, and a submerged current of relatively saltier water flows outward through the Strait of Gibraltar underneath the much larger inflow of surface water from the Atlantic.

Geologically, the Mediterranean was formed from the Tethys Ocean, a vast tropical sea, as the process of plate tectonics moved Africa northward toward Eurasia. As the continental plates began to collide during the Mesozoic era, a series of smaller plates split off, and the complex mountain ranges that almost surround the Mediterranean were folded upwards. Along with the folding, volcanic activity occurred and still continues more or less intermittently on Etna, Vesuvius, Thera (Santorini), and at many other places, including the almost continuously erupting volcanic island of Stromboli and its neighboring island of Vulcano, which along with its patron god, Vulcan, gave its name to the erupting mountain phenomenon. Sometime in the late Tertiary period, the continents joined at the eastern and western ends of the Mediterranean, and the sea slowly evaporated, leaving deep salt beds behind. This process may have occurred several times, because there are several of these salt layers. About five million years ago, the Atlantic Ocean found its way through the Strait of Gibraltar and filled the Mediterranean for the last time—so far, at least—in a gigantic waterfall that lasted a century or more. Ice Age glaciers did not invade the Mediterranean region except in the Alps and other high mountains, but the climate cooled, precipitation increased, and evaporation slowed. About twelve thousand years ago, the climate approached present conditions.

The oldest rocks in the Mediterranean are huge fragments of the ancient continental plates, dating back to the Precambrian era. But by far the most prevalent and extensive strata are of sedimentary limestone originally laid down under the Tethys Ocean, which has metamorphosed into fine marble in some places. Other formations, such as sandstone, shale, conglomerate, and coal seams, are found, along with ores containing metals such as gold, silver, copper, lead, and iron.

From surface rocks the process of soil formation began with the aid of vegetation; there are areas of highly fertile soil in parts of the Mediterranean basin, but generally speaking soils are thin and poor compared with northern Europe, for example. Because the predominant underlying rock is limestone, the soils tend to be calcareous. Because the climate is relatively dry, xeromorphic soils are prevalent. Because Mediterranean topography is generally mountainous, mountain soils that tend to be subject to erosion and relatively thinner and younger are important in almost every part of the basin. In most regions, the impact of human activity has disrupted the soil's structure. Even so, many Mediterranean soils, especially those in lowlands and coastal zones, are well suited to agriculture. Typical Mediterranean soils that form on limestone bedrock are *terra fusca* (black earth) and *terra rossa* (red earth). Terra fusca consists of sandy darker soils, usually brownish in color, that develop under forest cover and are not acidic because limestone is calcium carbonate and acts as a buffer against acidity. Terra rossa gains its red color from iron oxide that is present as an impurity in the limestone, and it tends to be rich in clay. It develops under maquis and garigue (scrubby evergreens and shrubs) and has been subject to more leaching by rainfall than has terra fusca. The moister soils of the north tend to be darker in color. In the Balkans and across the northern shores of the Black Sea there are chernozems, or black soils, rich in organic matter, that have developed under the cool grasslands. Southern France has well-watered soils of ochre coloration. In much of Spain, where the rocks are siliceous and crystalline, the soils have a fine sandy consistency. Along the desert margins on the south side of the Mediterranean, there are arid soils, lighter in color. In the absence of vegetative cover, the winds form dunes. Winds and rains carry away the lighter sand in many areas, leaving behind a desert pavement of rocks and pebbles. The opposite situation, namely saturation with water, occurs in the deltas of the large rivers such as the Nile, Po, Rhône, and Guadalquivir, where there is a heavy, silty soil and drainage is necessary if agriculture is to be practiced.

CLIMATE

Climate is a distinguishing environmental factor of the Mediterranean region. As Fernand Braudel understood it,

> That identical or near-identical worlds should be found on the borders of countries as far apart and in general terms as different as Greece, Spain, Italy, North Africa; that these worlds should live at the same rhythm; that men and goods should be able to move from one to another without any need for acclimatization: such living identity implies the living unity of the sea. It is a great deal more than a beautiful setting. (Braudel 1972, 231)

The ancient Greek medical writer Hippocrates held in his book *Airs, Waters, Places* that the climates of cities and other localities affect the health and psychological makeup of the humans who live in them.

The Mediterranean climate is intermediate between temperate and tropical, with two seasons: cool and moist from about October to April, and hot and dry in the other half of the year. The impression of the Mediterranean climate as a sunny one is undoubtedly true; the average number of hours of sunshine in Athens, Greece, is 2,655 per year, whereas Berlin, Germany, has only 1,614. The north and west have more precipitation and a longer winter than do the south and east. The preponderance of rainfall comes in the winter, with so little in the summer that many streams are intermittent. Watercourses in the drier eastern Mediterranean region are highly seasonal, with as much as 80 percent of annual flow limited to the winter months. For example, Nice in southern France has an average of 838 millimeters (33 inches) annual precipitation, but only 81 rainy days. Winter can be very stormy, raising the sea to heights that endanger shipping. Some of these storms sweep in from the Atlantic, but more typically they arise within the Mediterranean basin itself, especially in places like the Gulf of Lions in the western basin. But even during the wet season, rain is not constant and may be rare almost everywhere, although it can be very heavy during brief storms. On Malta, 279 millimeters (11 inches) of rain have been recorded in one day. The variation in precipitation is very high from year to year; one year may bring twice the average, the next year only half. Precipitation is also quite variable from place to place: The extremes are Port Said in arid Egypt, which averages

50 millimeters (2 inches) of rainfall; and Crikvenica on the mountainous Croatian coast, which receives an average of 4,626 millimeters (181 inches) of precipitation, some of it in the form of snow. In the drier eastern basin, typical amounts are Alexandria, Egypt, 217 millimeters (8 inches); Athens, 406 millimeters (16 inches); and Haifa, Israel, 679 millimeters (27 inches). In the comparatively rainier western basin, Marseille, France, receives 574 millimeters (22 inches); Algiers, Algeria, 765 millimeters (30 inches); and Rome, Italy, 923 millimeters (36 inches). The rainiest area is the eastern coast of the Adriatic Sea. The mountains may experience snow and frost in the winter and thunderstorms in the summer.

Temperatures are moderated by closeness to the sea in lowland areas. In many locales the difference between the warmest and coldest months is only about 14 degrees C (25 degrees F), and day may be a scant 10 degrees C (18 degrees F) warmer than night. There is no month in Palermo, Sicily, with an average temperature below 10 degrees C (50 degrees F). Near sea level, the temperature rarely falls below freezing. It seems incredible that Venice is as far north as Ottawa and that Thessalonica in Greece is further north than Denver. Snow may occur on one or two days in the winter in a locality such as Athens, but it ordinarily melts in a few hours. The Mediterranean winter can be unpleasant at times, however; storms often sweep the sea and raise waves to perilous heights. Sailors avoid venturing out from port during these months. Of course the *meltemi* winds of summer can also be dangerous; I have seen waves breaking over the bow of a cruise ship on which I was a passenger between Crete and Mykonos in July.

High temperatures can be oppressive in summer, especially in North Africa and the interior of the Near East. The dry, hot season in the Mediterranean lasts from May to October. In Greece, May Day is popularly called the "first day of summer." The months from June to September are rainless in much of the region. During these weeks the people favor a siesta in the afternoon, but they often complain that the heat keeps them awake both then and at night. The mean summer temperature in Naples is 33 degrees C (91 degrees F), and the nearby sea keeps the air humid. On the Libyan coast, temperatures rise above 49 degrees C (120 degrees F), and rain hardly ever falls. In more typical Mediterranean territory, especially near mountains, summer thunderstorms can break the drought and briefly cool the air.

On each of the eight sides of the Tower of the Winds, erected by Julius Caesar in Athens and still standing, is a relief of a personified wind, portrayed as if flying and bearing an attribute of the kind of weather it was observed to bring. For example, Kaikias, the northeast wind, carries a shield indicating its violence, whereas Zephyrus, the west wind, showers into the air a cornucopia of flowers. The pattern of winds varies in different parts of the Mediterranean, and

The Tower of the Winds, built in the marketplace of Athens by Julius Caesar, bore images of the eight winds. The two shown in the center are Notos, the south wind, who is emptying an urn to produce a shower, and Euros, the southeast wind, with a heavy mantle that predicts a violent storm. The tower also bears sundials and contained a rotating map of the sky that was driven by a water clock. (Photo courtesy of J. Donald Hughes)

they fluctuate from season to season. In the winter, a series of low-pressure centers form over the relatively warm Mediterranean Sea and move eastward. Depressions also move in from the Atlantic and are strengthened over the sea basins of the Mediterranean. The jet stream, which guides the lows, usually shifts into the Mediterranean, and during this period, moist winds from the west and southwest can bring rain. But the lows cause the warmed air within them to rise, drawing on winds from the surrounding lands. When this air comes from the high-pressure dome over the European continent, it enters as a series of cold, dry northerlies. Most infamous of these is the mistral (Latin *magister*, or "master wind"), which pours down the Rhône Valley, sometimes uprooting trees and blowing automobiles off roads. A cold wind called *bora* descends upon the Dalmatian coast. Similar winds are the *gregale*, which sweeps off the Balkan Peninsula across the Ionian Sea at times as far as Tunisia; and the *vardarac*, which comes down into the Aegean along the course of the Vardar

River. Cold winds also flow from Anatolia (central Turkey) toward the Aegean Sea, Black Sea, and the eastern Mediterranean. Lows over the Mediterranean also attract winds from the Sahara. These are hot, dry, and carry great quantities of red and yellow dust from the desert. The Italians call this type of wind *sirocco*, the Spaniards *leveche*, the Tunisians *chili*, and the Egyptians *khamsin* from the Arabic word for "fifty," because it supposedly blows for fifty days straight. This wind is extremely disagreeable, bringing sandstorms with dust that coats every surface, penetrates every crack, and makes food taste gritty. I have found dust from it inside a tightly closed suitcase. In the Maghreb it roars down the front of the Atlas Mountains as a *foehn* or katabatic wind (a dry wind that warms as it descends) and reminds one of a blast furnace. As the sirocco crosses the sea it absorbs moisture and becomes a dusty, muggy wind called *garbi* in Italy and along the Aegean. Clouds that are generated from this air mass often pour down rain that is red from the dust it contains.

In the summer, a mass of high pressure from the North Atlantic extends to central Europe, producing prevailingly northwest winds of continental dry air that move toward the African and Asian landmasses where heated air is rising. These winds are northeasterly in the Aegean, where they are called *etesian* (from *etos*, yearly) or *meltemi*. They make sailing easy in a southerly direction, but they also make it desirable to find harbors that are sheltered from the north. From this source Egypt enjoys north winds that cool Alexandria and make upstream sailing possible on the Nile.

PLANT LIFE

Mediterranean vegetation is more diverse than the plant life to the north, where glaciers covered the land for thousands of years, or to the south, where desiccation created the great Sahara in relatively recent times. The Mediterranean flora includes about 25,000 plant species. Greece has three times as many species of flowering plants as the British Isles, which are twice as large in area. More than a thousand species have been found by botanists within walking distance of Athens, and the same is true of Jerusalem. Many of the Mediterranean species are endangered. Of the 3,853 species endemic to individual Mediterranean countries (meaning they do not occur anywhere else), nearly 2,000 are rare or threatened. The plant communities of the Mediterranean region form a complicated mosaic affected by varying rainfall, elevation, exposure, and the impact of human activities. In general terms, they can be grouped into three zones according to elevation. The first two are forest zones; almost all parts of the Mediterranean lands originally supported forest.

A plume of smoke from a forest fire in Chalcidice near the Athos peninsula in Greece calls to mind the fact that fire is a common phenomenon and a formative force affecting the vegetation of the Mediterranean zone. (Photo courtesy of J. Donald Hughes)

The lowest zone, from sea level to between 650 and 1,000 meters (from 2,100 to 3,300 feet), displays types of vegetation most typical of the Mediterranean climate. In earlier times and in less disturbed sections this may be occupied by a forest of pines and evergreen oaks. Aleppo pines form extensive forests at lower elevations. There are thicker stands of water-loving species such as plane trees and willows near streams and marshes. In drier sections the trees are often widely spaced due to competition for moisture.

On dry hillsides periodically swept by fire, the plant community is a hardy, dense, drought-resistant brushland known by the French term *maquis* (in Spanish, *mattoral*, and in Italian, *macchia*), rarely more than 7 meters (23 feet) in height. Its species, including shrubby oaks, junipers, arbutus, and laurel, are adapted in various ways to survive fire or to return rapidly to burned areas. They tend to have broad leaves that are protected against moisture loss by thick, hard, hairy, leathery, or waxy outer layers. Forests, after removal, may be replaced by maquis, but maquis is also a natural and persistent association over a very large portion of the Mediterranean basin, and in many places it may represent the primeval vegetation. Maquis is not only drought resistant but also adapted to survive fire. After the periodic fires to which they are subjected, shrubs of the maquis establish themselves by sprouting from buried root crowns, as the kermes oak and bracken do, or from seeds that germinate in great heat or spread easily and take root in the bare, scorched topsoil. Some, like the cistus, strawberry tree, and thuja, can recuperate after a fire, often accelerating their growth at that time. Plants of the maquis are adapted to the long, dry Mediterranean summer. They have long root systems, and their osmotic pressure (the ability of a root to draw moisture from the soil) is high. They are evergreens, which enables them to take advantage of winter moisture and the growth it makes possible. Most of them bloom during the short spring, when the environment is briefly both warm and moist.

Maquis may also degenerate under repeated clearing, overgrazing, or fires to a low, tough community of typically spiny, aromatic shrubs that are often not as tall as the rocks among which they grow. This association is called *garigue* in French, *tomillares* in Spanish after the thyme that is one of its most notable species, and *phrygana* ("kindling") in Greek because the only wood it supplies is thin, short sticks. No doubt there was some garigue on the rockiest Mediterranean hillsides in prehistoric times, but people and their fires and herds have produced far more of it where there was formerly forest or maquis. Garigue plants include well-known spices such as basil, lavender, oregano, rosemary, thyme, and sage. The scents of garigue wafting out to sea give Mediterranean coasts the characteristic pleasant odors for which they are deservedly noted.

Where not even garigue can survive, there may be a grassland of species that flourish in the moist winter months, a Mediterranean "steppe." It contains many annual species that thrive in the moist half of the year and also root perennials and tuberous plants such as asphodel, one of its most characteristic components. In spring the asphodel steppe produces a colorful but brief display of flowers before the dry wind comes. The plants that survive best in this association are those most resistant to grazing, such as mullein, sea squill, and thistlelike composites. Members of the rockrose, legume, grass, mint, mustard, pink, buttercup, parsley, and lily families are common.

Above the zone just described, and up to 1,350 meters (4,400 feet) or so, if moisture permits, there may be a deciduous forest belt containing oaks, elm, beech, chestnut, ash, and hornbeams. This vegetation belt is called the Upper Mediterranean Zone. The deciduous tree communities are most often seen in the northern and western parts of the Mediterranean basin; elsewhere they are limited to moist enclaves or may not appear at all. They have suffered greatly from deforestation throughout history.

Still higher, up to the limit of tree growth, which varies from about 2,150 to 2,850 meters (7,000 to 9,350 feet) on the Mediterranean mountain ranges, the native vegetation in earlier times was a tall coniferous forest of pines, firs, cedars, and junipers interspersed with grassy meadows. The cedar forests, most famous in Lebanon, also occurred in the Taurus Mountains (in Turkey), on Cyprus, and along the Atlas Mountain ranges in North Africa. It was a favorite summer grazing ground for goatherds, whose goats were known to climb the branches to graze the foliage, and its tallest, straightest trees were sought by shipbuilders. Today this forest has almost disappeared from islands and from the highlands in Lebanon and the Atlas, and elsewhere its areas have shrunk due to logging.

Above tree line is a tundra of dwarfed flowering plants and lichens, below the rocky summits where snow lingers except during the late summers. There are glaciers in the barrier mountains, the Alps and Caucasus, although they are shrinking rapidly in the present regime of warming climate, but they do not persist on even the highest peaks of the Mediterranean proper such as Olympus and Etna. Tiny plants at these elevations are adapted to a brief summer growing season that makes it necessary for them to flower and set seed within a few short weeks before hard-freezing nights return. This high tundra zone is particularly subject to human damage, although it must be emphasized that all Mediterranean mountains are ecologically fragile.

The ecosystems of the deserts on the margins of the region, in the Sahara and the interior of Arabia, Syria, Iraq, and Turkey, are adapted to extreme conditions of aridity. Rainfall in some places is below 25 millimeters (one inch), and in the most arid spots, there may be no rain at all. No plants may be visible, with only bare rock and sand, in the driest locales; but in other places, plants with long roots may find underground water, or annuals may wait as seeds to take advantage of ephemeral rains. Here and there water rises to the surface and provides an oasis. The palm, so typical of oases today, is a human introduction of historical times.

The deserts are not totally devoid of plants, but those that grow are adapted to aridity. Some are annuals that spend most of their life cycles as seeds, and then when a rain comes—usually in winter—they explode in a brief paroxysm of growth, flowering, and fruiting. Others are low-growing perennials that have various ways of promoting their water intake, reducing water loss, and surviving the long periods of drought. Many have long root systems, especially those that grow in sand; the tamarisk can send its roots 30 meters (98 feet) underground, and the retama almost as far. During hot and dry periods many plants will reduce their transpiration by dropping leaves or failing to grow them. The ephedra has no leaves at all. Studies in the Middle Eastern deserts have shown that during the summer scarcity of moisture, desert vegetation will allow only a thirtieth of the evaporation from plant surfaces that may occur in the weeks after a spring rain. The adaptations of these plants are numerous, including thick cuticles and tissue walls, few evaporative stomata, small or inrolled leaves, water storage cells, and the ability to go into long periods of dormancy. Thorns, strong and unpalatable juices, and other devices that discourage browsing animals also favor survival in the desert ecosystems.

ANIMAL LIFE

The Mediterranean assemblage of animal life once included species that flourished in one or more of the vegetative zones previously described. The variety of animals was even greater than that of plants, especially if insects are included, as they must be. Despite their mobility, these components of the aforementioned Mediterranean ecosystems tend to be found in one or more of the life zones in broadly defined altitudinal bands. The Mediterranean forests and maquis were favorable habitats for wild animals. There are 1,050 species and varieties of land vertebrates, including freshwater fish. Just as humans have altered plant communities in the basin, so they have changed the distribution of animals by modifying their habitats, reducing their numbers, extirpating whole species, and introducing exotic species, domestic or wild.

The rich primeval Mediterranean fauna was related to that of the rest of Europe, with the addition of some animals typical of the African and Asiatic faunas. Many species are endemic, although the proportion is smaller than among the plants.

A number of wild mammalian species of the Mediterranean were herbivores that are relatives of domestic animals including goats, sheep, cattle, swine, donkeys, and horses. Other large herbivores such as bison and deer ranged the forest. The grassland margins of North Africa possessed a fauna resembling that of East Africa in modern times, with elephants, zebras, and many species of antelopes. The hippopotamus could be found along the Nile and other African and Asiatic rivers that flow into the Mediterranean, although it is missing from them nowadays. Smaller plant eaters were ever-present, including rabbits, hares, mice, voles, porcupines, and squirrels.

The next trophic level, as ecologists term it, consists of animals that eat other animals: the carnivores and insectivores. Larger predators then included lions, leopards, lynxes, hyenas, wolves, jackals, and foxes, all present on the European side of the sea as well as the Afro-Asiatic. Most ranged much farther in earlier times than later. There were smaller carnivores such as wildcats and weasels, and insectivores such as hedgehogs, shrews, and bats. Bears and Barbary apes were present as omnivores, eating both animal and vegetable foods. Large predators undoubtedly are the group of animals that has suffered the greatest proportion of reductions in range and numbers, and extinctions, almost entirely because of their persecution by humans.

Reptiles and amphibians were more prevalent in ancient times than today. There were several kinds of tortoises, both herbivorous and insectivorous. Snakes both poisonous and nonpoisonous preyed mostly on small animals. Small lizards such as the insectivorous geckos and chameleons, including one poisonous species, can be found. Crocodiles teemed along with hippopotami in rivers in Africa and Asia, and met the same fate; there are none near the Mediterranean at present. Amphibians, including frogs, toads, newts, and salamanders, as a rule were found in and near freshwater but today are seen more rarely, as is the case with other members of these groups around the planet, probably due to pollution that lowers their resistance to diseases. Frogs in ancient times formed choruses in marshes, which inspired the Greek playwright Aristophanes to include a chorus of singers imitating them in his comedy *The Frogs*, a satire of the tragedian Euripides.

The many species of birds in the region have been known by, and most of them eaten by, the Mediterranean peoples throughout history. Most are migratory in varying degrees, and there are some that leave the area entirely for part of the year, so that they are only seasonal parts of Mediterranean ecosystems. Some of them, like finches, pigeons, and sparrows, are seedeaters. Eagles, owls, hawks, and other raptors are carnivorous. Vultures, ravens, and magpies specialize in carrion but also eat the young of other birds and small mammals and reptiles. A number of birds are insectivores, and this makes them important to the ecological balance; to list only a few, there are swallows, thrushes, warblers, nightingales, starlings, and the crested hoopoe. There are summer visitors (hoopoe, swallows, orioles, warblers), winter visitors (rooks, short-eared owls, gulls, and other birds of the seashore), commuters between the northern and southern Mediterranean (avocets, ouzels, wrynecks), and year-round residents (ravens, kingfishers, tawny owls, buntings, wall creepers, and spotted woodpeckers). The rock dove became adapted to human buildings and has spread around the world as the common pigeon. It was raised in dovecotes and roasted as a delicacy, and its droppings were collected for fertilizer. Aristophanes's comedy *The Birds* demonstrates that knowledge of many species and their appearance and habits could be expected among Greek theater audiences.

The largest proportion of animals in the biological community, whether measured by number of species, number of individuals, or total biomass, consists of

Tortoises photographed in Delphi, Greece. The shells of tortoises were sometimes used by the makers of ancient lyres as sound boxes. Tortoises on Mount Parthenius in Arcadia, however, were considered sacred to the god Pan and were not killed. (Photo courtesy of J. Donald Hughes)

insects. They perform many functions in ecological processes. Many of them—from bees, beetles, butterflies, and moths to the musical cicada, cricket, and locusts—eat plants. Insects that consume animal material include praying mantises, wasps, hornets, and some beetles, as well as lice, fleas, flies, and mosquitoes. Other insects, such as the dung beetle, help in processes of decomposition. Various species of ants specialize in food sources; some are herbivorous, some carnivorous, and some practice mold agriculture or aphid pastoralism. I have observed ants from neighboring hills fighting a miniature war in the Mediterranean terrain near Delphi. Classical Greek and Latin writers, with good reason, described insects that like human blood, such as lice, bedbugs, fleas, flies, mosquitoes, and gnats.

Other herbivorous arthropods include the wood louse and millipede. Centipedes, spiders, and scorpions are primarily insectivorous. Snails and slugs, which are land mollusks, consume plants but serve as food for predators, even humans at times. Annelids such as earthworms also perform the helpful function of soil aeration and fertilization. The freshwater leech was mentioned by the physician Hippocrates, who advised its use for the removal of blood from

The color and remarkable crest of this bird make it a prominent part of the Mediterranean avian fauna, although it is found also in East Africa and India. One of the characters in Aristophanes's play The Birds *is a hoopoe. (Lanz Von Horsten; Gallo Images/Corbis)*

the back of the head, an inadvisable practice that continued on his authority as late as the twentieth century in less enlightened locales.

The Mediterranean Sea itself is the habitat for an aquatic community of life. Different temperatures, water depths, degrees of salinity, and forms of the seafloor provide a variety of habitats for aquatic ecosystems in the Mediterranean. Here life depends on food producers such as algae and phytoplankton, and on nutrients washed down from the land. Compared with the oceans, the Mediterranean Sea is less rich in marine species, whether calculated in terms of total number of species or in biomass (that is, the total weight of living organisms per unit of volume of seawater). This is the result of several factors. Its relative isolation meant that the Mediterranean could be entered only from the Atlantic through the bottleneck of the Strait of Gibraltar. The salinity of the inland sea, increasing to the east, discouraged some less adaptable organisms. Thus the western basin has a greater variety of species than the saltier eastern one. The temperature of the Mediterranean varies relatively little with depth, a condition not conducive to the cold vertical currents that favor the growth of plankton, the base of marine food chains. In general the Mediterranean is poor in plankton, and therefore its waters are noticeably clear and blue, or at least they were before the pollution that invaded them in the twentieth century. In saying that the Mediterranean is poorer in fish life than other seas, it should not be implied that fishermen of the coastlands found their work unrewarding. On the contrary, fish provided an important part of the diet of the region from ancient times to the present. Although fish are not as numerous as in the larger oceans, where temperature gradients and lower salinity are more favorable, more than five hundred species are found in the Mediterranean, along with algae, corals, shellfish, and sponges. Through most of history in the Mediterranean, fishing was an important economic activity, and although catches were usually destined for local markets, salted and dried fish became an important item of trade. More than a hundred species were economically important, from sharks and rays to eels, sardines, and anchovies. Flounder and sole preferred the relatively shallow sea bottom. Tyrian purple dye manufactured from the murex (rock whelk) was an important product from early times, providing the "royal purple" color (really more red than purple) for kings' robes and stripes on the togas of Roman senators. Large quantities of sponges, brought up by divers, were used in homes and industries.

Mammals such as whales, seals, and dolphins, all of which are predators of sea life, occurred in the Mediterranean waters abundantly in the past. Dolphins have long been regarded as friendly to people and have been credited with saving the lives of sailors such as the poet Arion. Many birds depended on the sea, whether they found their food on the shore (snipes, sandpipers), the surface (gulls, terns), or by diving under the surface (cormorants). Other seabirds included grebes, pelicans, and puffins. There were several species of sea turtles, which are marine reptiles and will be discussed in Chapter 6.

Invertebrate animals that live in saltwater are numerous, and some are sought as delicacies by fisherfolk. Some are crustaceans (barnacles, shrimp, prawns, lobsters, crabs); others are mollusks, including univalves (limpets, tritons), bivalves (oysters, mussels, clams), and cephalopods (squid, octopus, nautilus); echinoderms (starfish, urchins, sea cucumbers); and coelenterates (jellyfish, sea anemones, sponges, coral). Rivers and lakes offered freshwater habitats for ecosystems composed of many species. Fish in the lakes and streams included carp, perch, catfish, and eels. Fish such as salmon and sturgeon spent most of their lives in saltwater, but ascended rivers to spawn; the Nile River was a major ecosystem with many fish, and it supported a teeming population of aquatic birds, including ducks, geese, ibises, herons, and egrets. The hippopotamus ("river horse") and crocodile, now absent from Egypt proper, were mentioned earlier.

THE ARRIVAL OF HUMANS

Evidence of evolution leading to the appearance of modern humans exists in the Mediterranean basin, but it is fragmentary and tantalizing. The Fayum, a low area west of the Nile in Egypt, is the discovery place of a fossil anthropoid known as *Aegyptopithecus*, which lived sometime during the Miocene epoch (25 to 5 million years ago) and has been suggested as a common ancestor of apes and humans. Its appearance was monkeylike. Evidence of prototypic apes later in the Miocene, including *Dryopithecus*, has been found in France and elsewhere in Europe.

After that there is a gap of evidence in the region until *Homo erectus* ("upright man"), a species that many scientists believe to be a direct ancestor of modern humans, moved into the Mediterranean area from a probable place of origin in East Africa. This species had a cranial capacity of between 800 and 1,100 cubic centimeters, from about 60 to 75 percent of the average of modern humans, and a lifestyle including gathering, hunting of animals up to the size of elephants and rhinoceroses (which then still existed in much of the Mediterranean area), the use of fire, and the building of shelters. Stone artifacts such as hand axes suggest that *Homo erectus* came to Europe as long as 1.5 million years ago, but bone fossils date to a period between 500,000 and 200,000 years ago and show development in the direction of our species, *Homo sapiens*. Remains of *Homo erectus* have been found in Algeria, Morocco, Spain, southern France,

Greece, and Israel. The Petralona skull, found near Thessaloniki in northern Greece, has a cranial capacity of 1,230 cubic centimeters, dates to about 300,000 years ago, and has been described as an intermediate form between *Homo erectus* and Neanderthal humans (*Homo sapiens neanderthalensis*).

The Neanderthals were widespread in the Mediterranean basin from about 70,000 to 30,000 years ago, when the Ice Age was cooling the Mediterranean climate and swelling the mountain glaciers. Skeletal remains have turned up in Morocco, Spain, Gibraltar, France, Italy, Israel, and Iraq. Their brains were as large as those of modern humans, or larger, at about 1,400 cubic centimeters, but their faces were characterized by heavy brow ridges, sloping foreheads, and large noses. They were stocky and muscular, and with a finely shaped toolkit were excellent hunters of virtually every animal, large and small, in their environment. They lived in caves or shelters made of bones and skins and wore clothing of skins sewn together with bone needles and sinew. When they buried their dead, they painted the bodies with red paint made from iron oxide and manganese and made grave offerings of animal skulls, roasted meat, and flowers.

Humans of our species, *Homo sapiens sapiens*, spread out of East Africa through western Asia, across Central Asia, and then into the European Mediterranean lands. They reached Palestine perhaps 100,000 years ago, and Spain at the western end of the Mediterranean by 40,000 years ago. Sometimes called Cro-Magnons, they had replaced the Neanderthals by 28,000 years ago. Whether the Neanderthals interbred with modern humans and thus contributed to our ancestry is much debated. Genetic evidence suggests little or no mixing, but skeletons like that of a child found near Leiria, Portugal, and dated to 24,500 years ago show combinations of Neanderthal and modern traits that argue for miscegenation.

The new inhabitants of the Mediterranean made stone tools improved beyond those of the Neanderthals and were efficient hunters who undoubtedly reduced the numbers and ranges of large mammalian species. Their material cultures were those of the Palaeolithic (Old Stone Age). Some of them made convincing, energetic paintings of animals and humans on the walls and ceilings of caves. In the abundant ecosystems, they found plants to gather, fish and crustaceans to catch, and plentiful mammals to hunt. Sometimes too they were prey of the large predators such as the lion and cave bear. They used fire to cook, to keep warm, and to drive wild animals. Using materials such as wood, antlers, bone, and stone, they fabricated spears and spear-throwers, fishhooks, and eventually bows and arrows.

Technology is a series of adaptations to the natural environment, becoming more complex and powerful through time. It enabled early humans to make ever more far-reaching changes in the ecosystem, although it never freed them

from it. They depended intimately on the natural environment for their daily food, drink, clothing, and shelter. As a result of the undependability of supply of resources necessary to produce these items, the size of their groups was limited, and a natural balance, always somewhat precarious, was thus maintained between human population and the carrying capacity of the local environment. Individuals regarded themselves as integral members of communities, with the duty to provide the tribe with food, protect it against enemies, and seek power through visions, disciplines, and repetition of rituals. Elders received veneration because they incarnated the accumulated knowledge and wise judgment of the community. From the evidence of art, archaeology, and comparative ethnography it seems certain that practices supporting this balance were encouraged by oral traditions, including stories expressing views of the world, creation, the gods, animals, methods of hunting and gathering, and principles of family relationships, initiations, birth, and death. Ecologically considered, these traditions helped the community to adapt to the local environment and use it without destroying it. Peoples who share a hunting and gathering way of life tend to regard the world as animated by spirits, and to respect animals and plants as living beings endowed with power. Hunting, fishing, and gathering were carried out within rituals and surrounded by prohibitions developed through long generations of trial and error. Cave paintings represent shamanlike humans wearing animal skins, horns, and skulls, and dancing in movements evocative of the beasts. Stone, bone, and ivory carvings from this period probably represent a Lady or Lord of Wild Animals, a protecting power for the creatures, who would reward careful hunters and punish imprudent ones. Similar beliefs applied also to treatment of plants.

In spite of traditions that taught essential conservation, however, the Palaeolithic humans faced ecological crises, some of which they caused. The pressure of hunting drove many large animals to extinction, or perhaps acted together with climatic changes to hasten their disappearance. Some of the animals that vanished from the Mediterranean ecosystem by the end of the Ice Age were ones that humans might be expected to kill because they were predators on humans, such as the cave bear and hyena, or competed with humans in hunting other animals. Then there were giant herbivores like the mammoth and rhinoceros, sources of good quantities of meat, but eventually proving too unruly to domesticate. Fire was a great force employed by the hunters, who methodically set fire to forests, maquis, and grasslands in order to drive animals and to encourage the growth of new grass to feed grazing animals that were their quarry. But early humans could not control wildfire by any means except periodic burning to reduce its intensity when it occurred. Repeated fires could result in the replacement of forest with grassland. In ways like these, hunters

and gatherers made important impacts on ecosystems. If they seriously damaged the local ecosystem, however, they might have to leave or die. Over many generations, the groups that survived learned ways that tended to preserve the landscapes within which they could live well.

THE INVENTION OF AGRICULTURE

New ways of relating to the natural environment, involving the seasonal cultivation of food plants such as grain-bearing grasses and fiber plants such as flax, and bringing herd animals and their migrations under human direction, were adopted by many communities in the Mediterranean basin after the most recent Ice Age during a time of warming climate, broadly speaking from about 12,000 to 7,000 years ago. There is evidence that groups of people harvested wild grains near annually flooded lakes in the Nile Valley, in fertile pockets among the Syrian hills, and elsewhere between 14,500 and 13,000 years ago. Some of them, like the Natufians, used sickles with sharp stone teeth inset and other new tools. It was a natural but very important step from this to the deliberate saving and planting of seeds from one season to the next. When it was taken, that step made possible the feeding of more mouths in the community and the survival of larger, more sedentary populations in limited areas. Gradually humans moved into a new age, the Neolithic (New Stone Age), the age in which agriculture was practiced using well-made stone tools. The domestication of plants improved the dependability of the food supply and made a more concentrated population possible, but it also required a settled community to care for the growing crops. Along with agriculture came weaving, pottery making, and the fashioning of lighter, more sophisticated stone tools and weapons. People mined and traded obsidian for tools, the red dye stone hematite, and salt. The major crops were barley, wheat, oats, rye, legumes, and flax. The early planters used digging sticks and hoes, disturbing the soil to a relatively minor extent. They had quern stones and rubbing stones (metates and manos) for grinding grain prior to making it into bread. Farmers selected seed after the harvest for planting in the following year, and in this way new varieties of food crops evolved.

As important as the domestication of plants to Neolithic people was the domestication of animals. Sedentary farmers would learn to keep tamed species, but the original work of domestication was done by migrant people who, rather than following herds of grazing and browsing mammals just to hunt them, began with help from the already domesticated dog to protect them from predators and control their annual movements. Most herders were not nomadic wanderers, but practiced transhumance, that is, the alternate movement of

herds to higher summer and lower winter ranges. Pastoralism developed first in the Near East with goats and sheep, and later with cattle, pigs, and donkeys. These animals are adapted to ecotonal country—that is, locales where grassland, brushland, and forest interpenetrate—and thus found themselves at home in the Mediterranean landscape also favored by humans.

Herders and farmers retained the reverence formerly felt for wild species and extended it to the domestic animals and plants that had become the sustainers of human life. In rooms apparently dedicated to rituals in the agricultural village of Çatal Hüyük in Turkey around 6000 BC, for example, murals of bulls in black, white, and red decorated the walls; and the skulls and horns of bulls, covered by clay and painted with geometric designs, projected from the walls and floors. Domestic sheep and goats were represented along with wild leopards and vultures. The Mother Goddess, her arms raised in benediction, was shown enthroned between animals or in the posture of giving birth. This figure may be the "Lady of Animals" mentioned previously, or she may have symbolized grain, since more recent agricultural societies often gave grain the title of "Mother," for example, the Greek Demeter (*meter* is the Greek word for "mother").

There is little doubt that Neolithic people had an attitude of care for the earth and living creatures. But as their ability to change and control the natural environment increased, problems appeared. To open land for agriculture, they had to cut and burn forests and break up grassland sod. When farmers burned the vegetation, they found that the ashes temporarily enriched the ground. After a few crops and lessening harvests, they might let the land lie fallow while they cleared a new tract in a practice called swidden. Neolithic villagers' need for firewood and building materials depleted nearby forests. The removal of plant cover left slopes open to rainfall, which accelerated erosion, and consequently some hilly districts where early farmers practiced agriculture are now rocky and desiccated. The process of soil depletion was slow, and farmers who stayed permanently in the same area tried to find ways of countering it. On hillsides, they built terraces to reduce erosion. They discovered how to use manure and other fertilizers, and they planted legumes to enrich the soil for other crops. Neolithic farmers learned by trial and error, managing to remain in balance with the changing environment for long periods. Herders had some conservation practices, too, since if the vegetation in one area was reduced, they might be able to move their animals to other pastures, and their herds left manure to enrich the soil. But the fires herders set could cause damage, and overgrazing was a very serious and chronic problem.

Farmers and herdsfolk were well distributed around the Mediterranean by about 3000 BC and were creating openings in the forest cover for their activities.

Agricultural villagers and pastoralists watched the movements of sun, moon, and stars because the round of their labor was patterned by the seasons. They were conscious of their dependence on the Earth. Human numbers in the Neolithic period were relatively small compared with later times, even if the population was much larger than it had been in the Palaeolithic. People were making important changes in the vegetation and animal populations, but because their subsistence was always linked to the productivity of the local land and the fertility of their domestic flocks, they were led in considerable measure by necessity to maintain a balance with the ecosystems of which they were a part. Sometimes, however, the balance failed. Some groups began to experiment with new forms of agriculture and economic organization, as the next chapter will show.

References

Braudel, Fernand. 1972. *The Mediterranean and the Mediterranean World in the Age of Philip II*. New York: Harper and Row.

Semple, Ellen Churchill. 1931. *The Geography of the Mediterranean Region: Its Relation to Ancient History*. New York: Henry Holt and Company.

2

ENVIRONMENTAL IMPACTS
OF EARLY CIVILIZATIONS

We sit around our sea like . . . frogs around a pond.

—SOCRATES, QUOTED BY PLATO IN THE *PHAEDO*

Earth is a goddess who teaches justice to those who can learn, for the better she is served the more good things she gives in return.

—XENOPHON, *ECONOMICS*

If we evaluate the benefits of nature by the depravity of those who misuse them, there is nothing we have received that does not hurt us. You will find nothing, even of obvious usefulness, such that it does not change over into its opposite through man's fault.

—SENECA, *QUESTIONS ON NATURE*

Ancient history in the region including the Mediterranean basin and the Nile and Tigris-Euphrates valleys began with the invention of writing and the founding of the first cities during the fourth millennium BC (4000–3000 BC). It includes the periods of the city-states and empires of Mesopotamia, Egypt, Asia Minor, Persia, Greece, and Rome, to about AD 600, when the Roman Empire had fragmented into parts: various successor kingdoms, the western papacy, and the Eastern Roman or Byzantine Empire.

During this pivotal period, humankind initiated many new ways of relating to the natural environment. Cities, great centers of population and specialized occupations, appeared and altered the surrounding landscapes. Advances in agricultural techniques, such as the plow and extensive irrigation works, made increased food production possible. Methods of extracting and shaping metals, first copper and its alloy, bronze, and then iron, enabled the creation of more efficient tools for hunting, farming, and warfare. These inventions altered the face of the Earth. Humans transformed large areas from their natural state into busy scenes of production for human use. Fields and pastures expanded, while forests, wetlands, and other habitats felt the impacts of removal and drainage. Sometimes this transformation had a downside; evaporation of water used for irrigation, for example, could cause salt to accumulate in the soil and interfere with the growth of crops. Shrinking forests left room for farms but also left the ground open to erosion—and meant that timber and fuelwood might be less available. Hunting and habitat loss reduced the numbers of wildlife and extirpated some species in certain districts. Each of these environmental changes, and others, will be discussed in this chapter. Ancient environmental history in the Mediterranean basin and the Near East was a complex process. The natural environment in this region was far from uniform, and different districts were more or less affected by human settlement at various times and reacted to human impacts in different ways.

HUMAN SETTLEMENTS: THE URBAN REVOLUTION

The earliest cities appeared between 4000 and 3500 BC in Mesopotamia, and somewhat later in Egypt and Iran. In still later millennia, people founded other cities on islands and coastlands around the Mediterranean Sea. The city was a new form of human settlement, distinguished from the villages that preceded it not only by size of population, with tens of thousands of inhabitants, but also by the specialization of human occupations; the stratification of society into social classes; the increasing power of the state with its religious, political, and military institutions; and the emergence of aptitudes such as writing, the construction of monumental architecture, and the measurement of space and time. For the first time in history (but definitely not for the last time), powerful elite groups such as priests and royalty were able to direct the labor of the common people and the use of resources. The rationalization of this direction of their labor into tasks such as wall-building and canal-cleaning is contained in priestly writings such as the Sumerian *Myth of the Deluge:*

> *Of our houses, truly they will lay their bricks in pure places,*
> *The places of our decisions truly they will found in pure places. . . .*
> *He founded the five cities in . . . pure places,*
> *He established the cleaning of the small rivers . . .*
> *(translated by S. N. Kramer in Pritchard 1958, 28–29)*

The city also represented new ways of human relationship to nature. Agriculture became more intensive; for a human aggregation of such a large size, it was necessary for the labor of farmers to produce enough food to feed not only

The Acropolis of Athens takes advantage of a natural strong point in the landscape. It is defensible and dominates the surrounding country. (Roger Wood/Corbis)

themselves but also many others who did not work directly on the land. Inventions such as the ox-pulled plow and irrigation helped to make this possible.

Many cities built huge walls for defense against raids by nomads or the armies of other cities. These strong bulwarks look like symbols of a separation between the human-made environment of the urban center and the world of nature, cultivated or wild, outside the walls. This was true in a certain sense; an attitude of pride and dominance by city-dwellers over neighboring people and landscape is unmistakable in such ancient literary texts as the Sumerian *Epic of Gilgamesh:* "Gilgamesh went abroad in the world, but he met with none who could withstand his arms until he returned to Uruk" (Sandars 1972, 60). But in another sense the city depended on and indeed was part of an ecosystem that included the surrounding fields, pastures, forests, rocks, and waters. Cities reached even further outward to draw on the resources of near and distant lands and other cities through trade. Thus large cities such as Babylon, Athens, and Rome had impacts on the natural environment of places sometimes hundreds of

The aqueduct bridge of Segovia, Spain, shown where it enters the citadel, has 128 arches in two tiers, is 31 meters (102 feet) high at its tallest point, and the raised portion is about 825 meters (2,700 feet) long. Built of granite without mortar, it provided water until the mid-twentieth century. (Photo courtesy of J. Donald Hughes)

miles away. Babylon imported wood and stone from the Zagros Mountains and Lebanon, Athens secured a route to import grain from the northern shores of the Black Sea, and Rome made Egypt the granary of its empire.

One of the major needs of cities, especially in an area of generally low rainfall such as the Mediterranean basin, is the provision of a dependable water supply. At the start, the inhabitants might dig wells within the city itself, but a concentrated and growing population often rendered the supply inadequate, and wells within cities are notoriously subject to pollution. Local streams or rivers were initially used, but they might also become inadequate or polluted. These problems were addressed by the construction of aqueducts from sources at various distances near and far. The Assyrian king Sennacherib built massive aqueducts, fragments of which still stand. Hezekiah, king of Judah, built one that ran from a reservoir through a tunnel to supply Jerusalem. Aqueducts supplied many Greek cities; Eupalinus designed one in 530 BC that brought water to the city of Samos through a tunnel 1,000 meters (3,300 feet) long. Athens had an underground aqueduct running from springs on Mount Pentelicus with shafts for maintenance; the tunnel was still used in the twentieth century by the Athenian water department as a conduit for some of its water mains. Rome is famous for its numerous aqueducts, parts of several of them raised on long series of arches. These vital structures supplied fountains, pools, cisterns, and baths within the city. Roman aqueducts are surviving wonders of the ancient world, even in a nonfunctioning condition, and Frontinus, the Roman writer, boasted, "With such an array of indispensable structures carrying so many waters, compare, if you will, the idle Pyramids or the useless, though famous, works of the Greeks!" He was obviously wrong about the Greeks, since they had long before engaged in the construction of sophisticated aqueducts including innovations such as inverted siphons. By gathering water from springs, lakes, and streams over a large part of the countryside, hydraulic engineers made a major impact on the environment. The water that aqueducts took for urban use was for the most part no longer available to vegetation, wildlife, and agriculture. When all the aqueducts of Rome were fully operating, they carried a flow at least one-third greater than the average flow of the Tiber River through the city, almost 200 million gallons per year. The first Roman aqueduct, the Aqua Appia, built in 312 BC by Appius Claudius the Censor, ran entirely underground for ten miles. The Anio Vetus, constructed in 270 BC, crossed ravines on small bridges and was forty miles long. Later ones were even longer and were partly raised on high arcades that give the visual impression most modern people have of Roman aqueducts. The Aqua Claudia, built in AD 47 under the emperor Claudius, had an arcaded section of eight miles. Romans constructed aqueducts for cities other than Rome; Augustus ordered one for Alexandria, the Flumen Augusti. Other cities in the Roman Empire had similar if less impressive works. The emperor Hadrian built many aqueducts all over the empire. Moreover, when Theodosius II and his praetorian prefect Anthemius erected massive double walls to protect an expanded Constantinople in the early part of the fifth century AD, they provided the major aqueduct (built by Valens in the preceding century) with a cleverly constructed section to carry water under their new walls. Other emperors had provided the city with vast cisterns underneath the city streets to store water in case the aqueduct were to be broken by the enemy during a siege. One dating from the age of Constantine measures 60 meters by 70 meters, and its roof rests on 224 columns; a later one built by Justinian in the sixth century is even larger, with 336 columns, and still holds water today.

About 10 kilometers from Rome, the channels of five aqueducts crossed one another twice. The upper arcade carried the Claudian and Anio Novus aqueducts, and the lower arcade carried the Marcian, Tepulan, and Julian aqueducts. This reconstruction painting by Zeno Diemer is in the German Museum, Munich. (Bettmann/Corbis)

Ancient cities suffered from crowding, noise, air and water pollution, accumulation of wastes, plagues, and additional dangers to life and limb. There had to be some way of getting rid of excess water, sewage, and other wastes. In the earliest cities, and even in some later ones, these were simply discharged into the streets and other places, where they harbored insects, rats, and other animals that have adapted to living in human-built environments and often are carriers of disease. The city of Pompeii provided stepping-stones so that pedestrians could cross the streets without wading through muck, and to deal with the mess, many cities put street cleaners to work. Sewers began as open ditches and then were covered, at least partially. Athens had one that carried sewage to the fields outside the city where it could be used as manure. The greatest ancient sanitation engineers were Romans; the city of Rome had many sewers connecting with the *cloaca maxima*, or "main drain," running from the Forum down to the Tiber River. This was, however, an environmental liability during floods; when there was exceptionally high water in the Tiber, it was said that the drain in the center of the Pantheon, the great spherical temple in downtown Rome, looked like a fountain, but it must have been a malodorous one. Roman cities constructed public latrines, and the products of such facilities found use in tanning and bleaching.

Parks and gardens offered some relief from the crowding and pollution of ancient cities. There were parks in Mesopotamian and Egyptian cities. Queen Semiramis (Sammu-ramat) planted a paradise, or enclosed park, at Behistun, a mountain sacred to the weather god, and used a large spring to water it. Cimon, an Athenian general, sought popularity by throwing his large garden open to the people, making it effectively a public park. The idea of parks spread further to the west. Dionysius the Elder, a ruler of Syracuse in Sicily, planted plane trees in a park in Rhegium, a city in Italy that is within sight of Sicily. Alexander the Great and his successors planned cities that had open spaces and avenues planted with trees; Alexandria in Egypt is a good example of this. In papyrus documents from Egypt during the reigns of Alexander's successors, the Ptolemies, there is evidence for parks. It seems that virtually every village had *parádeisoi* (a Persian word for parks). One village, Cerceosiris, for example, had 4,700 *arourai* (about 3,000 acres) of parkland. The proportion of village land devoted to parks varied from 0.45 to 20 percent, possibly averaging around 10 percent; one town (Karanis) had at least 250 parks, but the size of most of them was quite small; few were more than two *arourai* (about 1.25 acres) in size. The Romans, in contrast, created extensive public gardens to improve the quality of life in the cities of their empire. Pompey, the Roman general, provided a large park in Rome itself with fountains, trees, and statues. Such amenities could preserve some of the elements of the natural environment that had disappeared elsewhere while offering people places for relaxation and amusement.

AGRICULTURE AND PASTORALISM

The economy of all ancient cities and empires rested on agriculture, and agriculture had far-reaching impacts on the natural environment. Domestic plants and animals produced almost all the food, as well as the fibers from which most clothing was manufactured. Agriculture consists of farming—that is, working the soil to produce crops and caring for farm animals. Pastoralism, or herding, involves tending flocks of grazing animals such as sheep, goats, pigs, cattle, horses, and camels; guiding their movements to places where they can find vegetation to eat; and gathering their wool, meat, skins, milk, and other products.

These two activities sometimes supplemented one another, but often they conflicted. The farmer and the cowherd are not always friends. A landowner might encourage herders to graze their animals on his fields of stubble after the harvest, because the droppings would provide organic fertilizer, but would not

Browsing in herds throughout the Mediterranean area, goats cause much degradation of the vegetation and contribute to deforestation by eating small trees, preventing the reestablishment of forest and maquis. Photographed on Samos. (Photo courtesy of J. Donald Hughes)

appreciate goats in the vineyards or cattle in the wheat. Pastoralists ranged widely, sometimes setting fires to encourage new growth to feed the flocks. Sheep ate the grass, and goats browsed bushes and trees, degrading the vegetation and opening the hillsides to erosion. However, the removal of vegetation also paradoxically reduced the danger of conflagrations.

The ancients observed that goats could damage plant cover. Plato knew how controversial the goat was, proposing an argument between a man who thought it a valuable animal and another who regarded it as a destructive nuisance. The comic poet Eupolis wrote a play with a chorus of goats and had them bleat out a list of their favorite foods:

> *We feed on all manner of shrubs, browsing on the tender shoots*
> *Of pine, ilex, and arbutus, and on spurge, clover, and fragrant*
> *Sage, and many-leaved bindweed as well, wild olive, and lentisk,*
> *And ash, fir, sea oak, ivy, and heather, willow, thorn, mullein,*
> *And asphodel, cistus, oak, thyme, and savory.*
> *(Macrobius* Satyricon *7.5–9)*

This could well serve as a botanical list of the most typical plants of the maquis, the Mediterranean scrub forest ecosystem, and it should be noted that a number of timber trees are also included on the goats' bill of fare. They were usually consumed while young and small. The significance of pastoralism is not that it destroys high forests but that it makes permanent what destruction went before. The effect of goats may be judged from the following statement: "In a place not far from Kopais we saw woody plants regenerating vigorously in a goat-proof enclosure, effectively demonstrating that the present sparse vegetation is due to grazing" (Greig and Turner 1974, 188).

Agriculture was invented as a subsistence activity in the Neolithic period, but it was transformed into a more widespread and intensive occupation in the ancient world. As it did so, the plow replaced the simple digging stick. In the Mediterranean and Near East, the scratch plow with a wooden plowshare drawn by animals, usually oxen, had been the dominant means of tilling the soil since early in ancient times, and it is still widely used today. One major side effect of it is erosion and soil loss, especially when it is used on sloping ground or hillsides. The most widely planted crops, wheat and barley, were the staff of life for the ancients in the various forms of bread and beer. Farmers also sowed and harvested many other annual plants, including legumes such as beans, peas, lentils, and chickpeas. There were also fodder crops and fiber crops: flax and hemp, for example. As human settlements expanded the areas devoted to agriculture, they cut down riverine forests, drained wetlands, plowed up grasslands, and altered the former habitats of wild animals.

Specialized crops for medical uses were also planted. One of the most renowned of these is the opium poppy (*Papaver somniferum*). Its place of origin as a domestic plant is unknown, and a truly wild population of it no longer exists, but it was grown by Neolithic villagers in the Alps in the fourth millennium BC and by the Sumerians in lower Mesopotamia not long afterwards. It spread throughout the Mediterranean lands in ancient times; the statue of a goddess wearing three heads of the opium poppy in her crown is known from late Bronze Age Crete (around 1500 BC). The uses of the juice to induce sleep and to calm pain, as well as the dangers of addiction, were well-known to physicians in Egypt, Assyria, Greece, and Rome, including Hippocrates, Theophrastus, Dioscorides, and Galen. The opium poppy must be planted on fresh, rich soil, because if planted for several years on the same soil the plants exhaust the elements necessary to their growth and degenerate as a result.

Because the early centers of agriculture were located in zones of meager rainfall, farmers depended on irrigation from major rivers such as the Tigris, Euphrates, and Nile and their tributaries. In Mesopotamia, the rivers were so undependable that societies had to build major works of irrigation, including extensive

Roman grain mills outside a bakery in Pompeii, Italy. The upper stone, formed as two intersecting cones, was revolved against the lower cone-shaped stone by slaves or animals. Grain was placed in the top and was ground between the stones until it emerged at the bottom as flour. (Photo courtesy of J. Donald Hughes)

systems of canals. The law code of Hammurabi, king of Babylon (1728–1686 BC), regulated the use of water in agriculture. In Egypt, where the Nile rose every year in a usually predictable way and flooded the fields, canals and water-lifting machines helped to extend the area of cultivation. Many historians believe that one of the reasons for the growth of ancient governments was to command the labor of large groups of people that was necessary to construct these works.

Environmental problems appeared as a result of these attempts to control nature, however. Canals carried not only water but sand, mud, and dissolved salt as well. Constant dredging was required to keep the canals flowing, and excess material often piled up along their banks until they were as much as 10 meters (30 feet) above the surrounding fields, presenting a danger in time of flood. As a challenge to sustainable agriculture, salt was just as important as silt. Irrigation in a hot, dry climate presents the threat of salinization. As water is spread across the croplands, it begins to evaporate, concentrating the dissolved salt. The remaining water may become brackish, and the fields often turn out to be crusty with salt crystals. Many plants cannot grow well, or at all, in such conditions, and the affected lands might have to be abandoned. Before that happened, crops more tolerant of salt might be planted for a while. For example, barley is more salt-tolerant than wheat, and there is evidence that barley gradually replaced wheat in parts of Mesopotamia, and also that yields generally declined, as they apparently did between 2400 and 1700 BC in Sumeria. In Mesopotamia, there appears to be a relationship between this kind of environmental degradation caused by humans and cultural decline.

In the lands to the north of the Mediterranean, including Greece and Italy, rainfall is relatively more abundant than in Mesopotamia and Egypt, but it comes mainly in the winter months, when heat and light energy for growth is least available. Farmers planted winter wheat and depended less on irrigation. The two most important agricultural products besides grain were olives, which are a tree crop, and wine from perennial grapevines. Both of these bear fruit in the autumn and were produced earlier in Egypt and the Near East. Wine in particular was being made, preserved with resin, and kept in clay jars by 5400 BC in the Zagros Mountains of Iran.

Market gardens, whose fresh vegetable produce invited timely sale, were located close to cities, or even within them. Ancient authors complained about declining yields at harvesttime and speculated that the soil was losing its fertility. But Greek and Roman farmers were quite aware of ways to restore their farms by applying fertilizers and manure, by composting and crop rotation, and by planting legumes that enrich the soil. The Greek writer Xenophon summarized this by saying, "Earth willingly teaches justice to those who can learn; for the better she is treated, the more good things she gives in return" (*Economics* 5.12). An

A vineyard in May, near Pisa, Italy. Vines and their fermented product, wine, formed one of the three staple Mediterranean foods throughout recorded history, along with grains and the olive. (Photo courtesy of J. Donald Hughes)

author from the Platonic corpus remarked, "It is thus that Earth conceives and yields her harvest so that food is provided for all the creatures, if wind and rains are neither unseasonable nor excessive, but if anything goes amiss in the matter, it is not deity we should charge with the fault, but humanity, who have not ordered their life aright" (Plato *Epinomis* 979 A–B). The decline of ancient agriculture and the loss of soil fertility, therefore, were not due to ignorance. Other factors also interfered with agriculture. When vegetation is removed from the land, particularly on slopes, soil erosion can become severe. Farmers countered this by building stone terraces to hold the soil in place. But the policies of ancient governments were all too often preoccupied with war and other nonagricultural matters. Taxes such as the Roman *annona militaris* (an annual tax to support the army) bore heavily on the agricultural sector, depriving farmers of resources they could have used to improve the land. Citizen farmers were conscripted for military service, so they were forced to be absent from their land and were sometimes killed in battle. Often war devastated swaths of land; armies waged deliberate environmental warfare, chopping down olive trees and burning farmhouses. A biblical commandment forbids Jewish soldiers to cut down fruit trees while besieging a city (Deuteronomy 20:19–20). This indicates that such destruction was a common practice of armies. It is no wonder that ancient writers complained of abandoned fields (*agri deserti* in Latin). Furthermore, when terraced hillsides were abandoned, the terrace walls were no longer maintained, and when they collapsed the amount of erosive material that washed down into lowlands and coasts greatly increased. This is one of the reasons why harbors silted up in the war-ravaged decades of the later Roman Empire.

In Italy and some other Roman dominions during late Republican times (200 to 27 BC) and throughout the Roman Imperial period, affluent landowners converted extensive holdings, called *latifundia* (ranches), from cropland to pasture for herds of sheep and cattle, which they found more lucrative. Livestock ranches could be run profitably with much less labor than would have been required from crops, and that labor could be assigned to slaves. The government compensated for declining grain production in Italy by imports from northern Africa, including Egypt, and lands around the Black Sea. From the late second century BC onward, much of this grain was distributed at subsidized prices to common citizens in the city of Rome. This demand must have encouraged farmers in the exporting regions to expand the cultivated area into marginal, environmentally vulnerable lands. In Egypt and North Africa generally, these would have been on desert fringes open to desiccation during droughts, where removal of vegetation may have begun a disastrous process of desertification.

HUNTING, FISHING, AND BIODIVERSITY

Hunting did not end with the invention of agriculture. It continued as a way of supplementing food supplies; of procuring skins, furs, and feathers; and of killing wild animals that preyed on domestic herds. As a young shepherd, David killed lions and bears that attacked the sheep he guarded, a practice that prepared him to kill the giant Goliath with a slingshot (1 Samuel 17:34–36). Jewish dietary laws prohibited the slaughter of animals for food unless they were in captivity, however, which prevented hunting from becoming a part of traditional Jewish culture.

Kings of many nations engaged in hunting for recreation and as a form of royal propaganda. Egyptian reliefs show the pharaoh hunting wild bulls, and the king of Assyria is often shown killing lions, sometimes with a sword that goes all the way through the beast and emerges from its back. Kings might have tried to reserve lions as exclusive royal prey, but this does not seem to have kept shepherds from defending their flocks. Egyptian paintings show nobles hunting birds in marshes, accompanied by cats that apparently served as retrievers and were rewarded by meals of fish. Other works of art show dogs assisting in hunts for antelope in the desert. Greek and Roman writers considered hunting to be an appropriate sport for upper-class men, also assisted by trained dogs, and wrote manuals telling how to conduct it properly.

Commercial hunters not only supplied venison and birds for city markets but also scoured the mountains for beaver pelts and skins for various kinds of leather. Trade brought wild animal products from great distances. Ivory from African and Indian elephants was used as a material in works of art, including furniture, utensils, and colossal statues such as the chryselephantine image of Zeus at Olympia—one of the Seven Wonders of the World—and the equally impressive figure of Athena in the Parthenon at Athens. Leopard skins, ostrich eggs and feathers, and countless other items were carried northward from the interior of Africa by boat down the Nile or by camel along the trans-Saharan trails. The pelts of fur-bearing mammals of the Alps and northern Europe were eagerly sought. Ancient armies pressed elephants into service as living "tanks" in ancient India, and the practice spread westward from the time of Alexander the Great onward. African as well as Indian elephants proved amenable to military training and performed other tasks.

Animals formed a source of entertainment from early times. Alexandria had a zoo and botanical garden connected with its great library. A fantastic procession of exotic animals was held in that city in the decade of 270 BC by Ptolemy II Philadelphus in honor of the return from India of his predecessor, Alexander the Great, personified as the god Dionysus. It included many beasts and humans from India as well as the parts of Africa south of Egypt, and chariots or carts drawn by elephants, goats, saiga antelopes, oryxes, hartebeests, ostriches, deer, and onagers, as well as the more usual horses, camels, and mules (Rice 1983, 16–19). Many other species marched in the parade or were exhibited in cages, including a white bear. Modern scholars argue as to whether the latter was a polar bear or merely an albino brown bear, but a Roman emperor later showed a white bear in a pool catching seals; that one must have been a polar bear (Jennison 1937, 7, 70–71, 189; Calpurnius *Eclogues* 7.43–46).

Ptolemy II had inherited from his father, Ptolemy I, a collection of wild animals in captivity near the palace in Alexandria, which was no doubt the source for many of the beasts in the procession (Diodorus Siculus *Historical Library* 3.36.3). Among other species Ptolemy II obtained for this collection were a chimpanzee and a gigantic python. This renowned menagerie was maintained by his successors. To collect such animals for study, the Greco-Macedonian rulers of Egypt sent out far-ranging expeditions, some of which penetrated into Ethiopia by way of the Red Sea. In 46 BC, Julius Caesar brought a giraffe to Rome, the first one ever seen there, which was probably a gift to him by Cleopatra from the royal zoological gardens at Alexandria (Pliny *Natural History* 8.18 (27) 69; Dio Cassius *Roman History* 43.23.1–2).

A mosaic from the Roman Villa of Casale, near Piazza Armerina in Sicily. This image from the "Corridor of the Great Hunt" displays the leading of a captured African elephant aboard a boat for transportation to Italy for use in a venation or show hunt in an amphitheater. The mosaic dates from the early fourth century AD. (Photo courtesy of J. Donald Hughes)

Trained baboons and bears appear in ancient art. Crowds gathered for dogfights and quail fights, and cockfighting was a spectacle as soon as domesticated Asian jungle fowls, the ancestors of farmyard chickens, had been introduced to the Mediterranean area from India. But nothing matched the Roman amphitheater for sheer spectacle and mass destruction of animals and human life as well. Unarmed criminals were exposed to wild beasts that had been starved or goaded into attacking them. The floor of the amphitheater was called *arena*, the Latin word for the sand that covered it in order to soak up the blood from animals and humans. Mock hunts called *venationes* formed a popular part of the shows. Sometimes the animals were exhibited in clever ways, but more often they were killed. Leopards, bears, lions, and elephants perished in the extravaganzas, along with crocodiles and hippopotami from the Nile. Julius Caesar exhibited lynxes from Gaul. Nero displayed polar bears catching seals. The number of animals destroyed in these spectacles was excessive: Augustus held twenty-six venationes in which 3,500 beasts were killed, including tigers from India. Titus dedicated the Colosseum by slaughtering 9,000 animals in one hundred days, and Trajan celebrated his conquest of Dacia (Romania) by taking the lives of 11,000 wild creatures. An organized business for the capture and transportation of animals was required to satisfy the demand. It was a difficult job, not least because the beasts, many of them quite large, had to be kept in good condition in cages and boxes as they were transported from place to place. It seems certain that more animals died en route than reached their destinations. Once they reached the cities where they were to appear in the amphitheaters, they had to be kept in menageries and fed until needed. A large facility for elephants existed near Rome at Ardea under the supervision of an official who bore the title *procurator ad elephantos*. Few people raised any objection to the sadistic displays. Among those who did was Cicero, who maintained that elephants have a kind of fellowship with humans. Both Cicero, when governor of Cilicia, and King Ptolemy of Mauretania refused to let their people capture and ship animals for the arena. Marcus Aurelius, the Roman philosopher and emperor, disliked the cruelty of the games but did nothing to stop them. The Pythagoreans and the writer Varro objected to hunting per se.

Populations of wildlife diminished as a result of all these activities, and various species became extinct in one area after another. Egyptian pharaohs hunted elephants in the Orontes Valley in Syria as late as the fifteenth century BC; it is not known when the pachyderms became extirpated there. Lions roamed Greece at the time of the Persian Wars; Herodotus reports that they attacked the camels in the Persian baggage trains. They had disappeared by the first century BC. The Greeks also extirpated leopards and hyenas. Hunting eliminated wild cattle, sheep, and goats from some of the islands. Sacred groves offered protection to some species, but they were limited in size. The Romans decimated the wildlife of North Africa, where elephants, rhinoceroses, and zebras eventually could no longer be found. They hunted out the tigers of Armenia and northern Iran. Bird populations also suffered.

The Mediterranean supported a large fishing industry, although sea life was never as rich there as in the Atlantic and other oceans. Local fisherman brought fresh fish to market almost every day, and at times entrepreneurs imported salted fish from Egypt, the Black Sea, and the Atlantic coast of Spain. They also cultured fish in saltwater and freshwater ponds. Sponges and shellfish, including oysters and murex, the source of the famous Phoenician purple dye, were collected. Like hunting, fishing was a popular sport celebrated in literature. There were complaints, however, of the depletion of fisheries. The exorbitant prices for some trendy species on the Roman market may indicate that there were fewer left to catch than before. An example of the intentional introduction of a fish comes from Roman times. Optatus Elipertius, prefect of the fleet under the emperor Tiberius, collected numbers of a species of fish, the brilliantly colored *scarus* (parrot wrasse), from the seas between Rhodes and Crete and planted them along the western shores of Italy. Regulations forbade catching this species for a period of five years, constituting as a result one of very few surviving examples of legislation intended to protect a species. This is reported by Pliny the Elder, who states, "Careful protection by land and sea rendered poaching almost impossible. For the period of five years any scarus caught in the nets had, under heavy penalties, to be returned straightway to the water. The enforcement of these wise regulations effected [a] mighty thriving of the fish . . ." (Pliny *Natural History* 9.79). Sergius Orata planted oysters in the Lucrine Lake, and others attempted to establish oyster beds on the island of Chios and near Bordeaux (Aristotle *Generation of Animals* 3.11). Wealthy Romans kept fishponds, seeking out and paying well for exotic fish distinguished by their beauty or tasty flesh. The finest rare fish might have sold for their weight in gold; three mullets once brought 30,000 sesterces at Rome, and Pliny writes that this species rarely exceeded two pounds in weight (Suetonius *Tiberius* 34; Pliny *Natural History* 9.30).

FORESTRY AND DEFORESTATION

Loss of forests has been one of the most notable and widespread effects of human activity on the natural environment in world history, and the area of the Mediterranean and Near East provides one of the earliest examples of this process. Evidence today increasingly demonstrates that significant deforestation occurred during the ancient period under consideration in this chapter. Forests did not disappear from the entire Mediterranean basin, but their area was sharply reduced, and as a result there was increased erosion, local desiccation of climate, loss of habitat for some species, and decreased availability of forest resources such as timber for construction and wood for fuel. As Jared Diamond describes it,

> With the tree and grass cover removed, erosion proceeded and valleys silted up, while irrigation agriculture in the low-rainfall environment led to salt accumulation. . . . Thus, Fertile Crescent and eastern Mediterranean societies had the misfortune to arise in an ecologically fragile environment. They committed ecological suicide by destroying their own resource base. (Diamond 1997, 312)

Early literature and art illustrate the human use of forests. The *Epic of Gilgamesh* portrays its hero, the king of Uruk, and his companion Enkidu undertaking an adventurous journey in search of a cedar forest, which they cut down after slaying its guardian Humbaba, a creature that combined divine and natural characteristics. This was during the Bronze Age, when demands on forests escalated due to increasing use of wood in more densely populated cities, where it was needed for buildings (even clay brick structures needed beams for their roofs) and as fuel. Wood was sometimes directly burned for fuel, but more often it was processed into charcoal first by partial combustion. There were few forests in the valley of Mesopotamia, and most of those that formerly existed had been removed to make room for agriculture, so any city that wanted building timber had to get it from distant mountains in Iran, Lebanon, and elsewhere. These forests suffered in turn as the Assyrian and Persian empires exploited them. There is archaeological evidence of severe deforestation in the early Bronze Age in the Danube Valley on the northern edge of the Mediterranean basin, where lack of fuel may have weakened a promising cultural development (Tringham 1971, passim). In southern Greece and the islands, evidence from ancient pollen indicates that the species composition of forests changed significantly during the Bronze Age, evidently as a result of human activity, since the trees that increased include those favored by humans, such as olive and plane.

Classical writers often refer to forests that had disappeared due to human actions in the past, down to and including their own times. Plato wrote a deservedly famous passage in the *Critias* (111 B–D) describing the deforestation of the mountains of Attica, his homeland. He says that the former existence of forests there can be demonstrated by the fact that beams taken from those mountains could still be seen in buildings in Athens. The forests had not regrown; where they had been, he observes, there was in his own time only "food for bees," that is, low-growing plants and shrubs. After the removal of the trees, the rains had washed away the exposed soil, leaving mainly rocks "like the skeleton of a man wasted by disease, all the fat and soft earth having been washed away, and only the bare bones of the land being left." Formerly the forests had absorbed the rains, storing the water and releasing it in springs, but in Plato's own day the former springs, marked by shrines to their tutelary deities, were dry. Theophrastus reported that in his time (the fourth century BC), wood of good quality, especially large timbers needed for temple rafters and ships' masts, had been used up near Athens and had to be sought in less accessible places such as the mountains of Macedonia (*Enquiry into Plants* 3.2.4,6; 3.3.2; 4.5.5). Roman authors such as Varro and Pliny the Elder made similar comments. Diodorus chronicled the passing of the rich forests of Spain and his homeland, Sicily. Livy writes that the Ciminian Forest north of Rome had been a barrier to Fabius's army some two centuries before his time (the first century BC) but had since disappeared. Strabo lamented that the forests near Pisa had good shipbuilding timber that was being consumed while he was writing for buildings in Rome and countryside villas of Persian magnificence (5.2.5). Lucretius described the process:

> [Folk] made the woods climb higher up the mountains,
> Yielding the lowlands to be tilled and tended.
> (De Rerum Naturae 5.1370–1371)

The most important consumption of wood in the ancient Mediterranean, perhaps as much as 90 percent, was as fuel for lighting, for cooking, for heating public facilities such as baths, and in industries such as ceramics and metallurgy. Areas around mining and smelting centers became among the most deforested. Towns and cities required the services of woodcutters, charcoal burners, and haulers who brought fuels to marketplaces. A second important use was timber for building. Much of this trade was carried on by water; logs were floated down rivers to timber ports and loaded onto merchant ships. Rome received timber on the Tiber River through a special port and possessed a lumber market. Shipbuilding was a third major use of wood in war and peace, the one most often mentioned by ancient writers. Faced with a declining supply of large timber for purposes such as these, the emperor Hadrian established a forest reserve on the mountains of Lebanon where trees of the most important species were declared to be the property of the emperor and could not be cut without his permission. More than a hundred stone boundary markers remained in place until the twentieth century, inscribed with warnings against timber thieves:

> Boundary of the forests of the emperor Hadrian Augustus:
> Four species of trees reserved under the imperial privilege.
> (Meiggs 1982, 85–86, retranslated)

It is not known just which species were included in the four that the emperor protected, but it seems certain that the famous cedar of Lebanon (Cedrus libani) was one of them.

Cutting trees in a forest will not result in deforestation unless, speaking in general terms, the amount of wood removed is greater than the amount the forest provides through reproduction and growth. But in the ancient Mediterranean and Near East, the withdrawals from the forest account in terms of biomass often exceeded the income. One factor in consumption was constant grazing by animals, especially goats, which eat the young trees growing from seed before they have the chance to mature. Early and Classical use of forests for charcoal and other industrial uses stripped forests, reducing the vegetation in many areas to maquis (brushland).

When large, old trees are cut for timber, the character of the forest changes considerably. Not all trees die when they are felled; some species will sprout from the base and produce new wood or "coppice." But the Mediterranean has a relatively small area available for the exploitation of this kind of wood. The dominant trees for timber are conifers, especially pines, which will not sprout after they are cut. Removal of forest for agriculture and for settlements transformed large areas of the Mediterranean landscape.

Fires, whether those started deliberately by humans or wildfires kindled by lightning or volcanic eruptions, have long been common in the Mediterranean forests, especially during the long, dry summer. They expose the hillsides to erosion. But Mediterranean vegetation is generally adapted to fire and recovers if not burned too often and if the new growth is not heavily grazed. Maquis in particular is a fire-adapted plant association that recovers after burning because various species have means of surviving through resprouting or seeds that are fire-resistant or that spread quickly into burned areas on the winds. When combined with other factors, however, fire can contribute to deforestation. Roman writers knew that the destruction by flock herders included setting fires to burn woody vegetation and encourage the growth of grass. Virgil described it:

> Just as in summer, when the winds he wished for
> Awake at last, a shepherd scatters fires
> Across the forests . . . (Aeneid 10.405–409)

These fires, and wildfires started by lightning or volcanoes, usually burned until they reached a barrier or were put out by rains; they were fought only if they threatened a settlement.

Fire, much of it human-caused, determines the characteristic patterns of Mediterranean vegetation. As Stephen Pyne observes (1997, 85),

> If, in fact, it were possible to remove fire completely, many typical Mediterranean species would slide into insignificance, perhaps oblivion, and many common communities would metamorphose as dramatically as Kafka's Gregor Samsa [who turned into a large insect]. They would resemble those sacred groves from which fire, along with other disturbances, has been excluded and the genetic potential of the Mediterranean biota has blossomed into distinctive, often unique forms. The cedar groves of Lebanon resemble nothing else in the degraded landscapes of the Levant. The protected forests of Mount Athos differ from Thrace as much as modern Athens does from ancient Argos. The monastic groves on the Saint-Baume massif rise like an apparition from the maquis of Languedoc.

Sacred groves as such are discussed in the section on conservation in this chapter.

One of the most important sources of evidence for historic patterns of vegetation is palynology, the study of deposits containing pollen that were laid down in the past in places such as lake bottoms, caves, or other protected spots. Pollen survives remarkably well, and the pollen grains from different plant species usually have distinct patterns that can be distinguished readily. By counting the frequency of pollen grains in such deposits, an idea of the species composition of past vegetation can be gained. The deposits can be dated by the radiocarbon method. The pattern of changes in vegetation revealed for ancient times is complex, but it can be said that in many areas there appear to have been cycles of deforestation and recovery. Forests may have become established and then been removed many times. In northern Greece, for example, palaeobotanists discovered a pattern in the pollen evidence indicating that mountain forests survived best in peaceful times, but when invasions occurred, peasants moved from the lowlands into refuge areas in the mountains, where they cleared forests in order to plant wheat and barley. When more stable conditions returned, they abandoned these retreats and moved back to the richer plains, allowing forests at higher elevations to recover. Because invasive movements of peoples occurred often over the centuries in Macedonia, this cycle was repeated several times there. Palynology also shows that forests persisted in parts of the north down through medieval times, whereas they were gone in some populated areas as early as the Bronze Age. For example, pollen cores show that pinewoods had disappeared from some coastal areas near Pylos at the southwestern corner of Greece by the Middle Bronze Age or early in the Late Bronze Age.

Deforestation around the Mediterranean Sea in ancient times increased the danger of flooding and erosion. Forests absorb rainfall and release it slowly, but when they are removed the water rushes downhill with little to hold it back. Cities below denuded hillsides were likely to experience floods, and Rome, located on the banks of the Tiber River, whose watershed was subjected to progressive deforestation, had floods of increasing frequency and severity. The first recorded flood in the history of the city of Rome occurred in 241 BC. There was also a catastrophic flood in 54 BC, which Dio Cassius attributed to heavy rainfall or to the sea driving back the river water. Undoubtedly the rainfall was a cause, but Dio missed the fact that if forest had still been present in the Tiber drainage, the severity of the flood would have been greatly reduced. Studies have shown that areas covered with the Aleppo pine so common on the Mediterranean hillsides will hold the water and soil more than fifty times as well as denuded slopes. A flood lasting three days in 23 BC made the streets of Rome passable by boats, and there was a flood in the next year as well. Floods occurred in AD 15 in the reign of Tiberius, and also under the emperors Vespasian, Nerva, Trajan, Marcus Aurelius, and Caracalla. There must have been floods that did not receive mention in the historical records as well. Rain, unimpeded by vegetation, washes the soil away from mountains and deposits it in lowlands and in the sea along the coasts where streams and rivers come down. Siltation clogged ancient harbors. The Romans had to dredge new basins for shipping several times at the port of Ostia. Sometimes, as at Miletus and Ravenna, a seaport became isolated from the sea. Elsewhere, marshy areas expanded along the coasts, providing new habitats for wildlife but also breeding grounds for malaria-carrying mosquitoes. People avoided settling in such places because of the disease, although they did not understand how it was transmitted. Deforestation could change the local climate, making it more arid due to loss of transpiration, and windier, because the trees that had slowed the movement of surface air were missing. Theophrastus personally observed and recorded changes in local climates: after people cut down the trees around Philippi, for example, the waters dried up and the weather became warmer (*Causes of Plants* 5.14.5). Such impacts were no doubt most serious in marginal areas such as the edges of the Sahara and Arabian deserts. An economic effect of deforestation was to make wood scarcer and therefore more expensive. Ancient inscriptions and other evidence support the thesis that there was a general trend of rising prices for timber during Classical times (Meiggs 1982).

EXTRACTIVE INDUSTRIES

Industrial technology was not as important a segment of the total economy in ancient times as it is in the modern world, but it advanced in magnitude and techniques and was able to make many changes in the environment. Some industries involve direct removal of materials from the earth. Mining, quarrying, and digging of substances that are raw materials for pottery, glass, bricks, concrete, mortar, and fertilizer all impact the landscape, producing pits and tunnels along with scars that expose the ground to erosion and accelerate leaching of chemicals into water. These and associated processing industries such as metallurgy and ceramics place demands on forests for wood for construction, and wood and charcoal for fuel.

In early societies, including Mesopotamia and Egypt, the majority of construction consisted of clay bricks. When these were fired, they required vast amounts of fuel. Even when they were simply dried in the sun, they represented a loss to arable soil. Egypt found plenty of stone in the cliffs that flank the upper Nile for monumental construction, but ordinary people lived in mud-brick structures. With the advent of the Bronze Age, these early civilizations learned how to mine and process the ores of a variety of metals and exploited them, developing new methods as they did so. Copper was rare in alluvial deposits, so the Mesopotamians sought ore in the northern and eastern mountains, and the

Egyptians found copper mines in Sinai. Tin, the other component of the alloy bronze, was much rarer. Eventually the Phoenicians brought it from as far away as Cornwall in Britain.

Mining extended to virtually every area of the ancient Mediterranean basin. Both Greeks and Romans looked down on mining as degrading labor, though they were willing to profit by it. The Athenians struck their famous coins, the drachmas bearing the image of an owl, from silver mined at Laurium. Silver mines on the island of Thasos had been worked by the Phoenicians before the Greeks. The Iron Age began about 1000 BC, sparking a search for exploitable ores of that metal. The Spartans had an iron mine in southern Laconia, the source of fine metal that they manufactured into weapons as well as the cumbersome Spartan iron money called obols. The Etruscans became skilled ironworkers, and evidence of that is found in many archaeological sites in Italy. Philip of Macedon conquered Mount Pangaeus in order to exploit the placer gold there, thus gaining the riches he needed to hire spies and to bribe corrupt officials. Spain was a Roman province whose mines and smelters were famous. Greek and Roman miners accomplished work on an amazing scale considering the level of their technology. Many mines were of considerable size even by modern standards. At Laurium, more than two thousand vertical shafts gave access to over eighty-seven miles of tunnels. Some Roman silver mines in Spain were as much as 250 meters (800 feet) deep. The methods used for extraction of the ores were washing them from the surrounding material by placer and hydraulic mining, open-pit mining, and tunneling into veins deep below the earth.

A never-ending problem in mines was drainage, which along with the need for air supply limited the depth to which shafts and tunnels could be sunk. Where topography permitted, miners could dig tunnels as drains. Elsewhere, water was raised by bailing, or by more sophisticated means such as the Archimedean screw pump, piston pumps, or waterwheels. A series of eight waterwheels was found in a mine at Rio Tinto, Spain. Drainage from mines polluted water with many substances, some poisonous.

Quarrying presented many problems similar to those encountered in mining. Since large, potentially useful blocks of various kinds of stone had to be cut and removed, most ancient quarries were of open-pit type, but sometimes tunnels and galleries were excavated underground. In the quarry at Syracuse where Athenian soldiers were imprisoned after their defeat in the Sicilian Campaign, there is a single quarry face 27 meters (88 feet) high and 2 kilometers (1.2 miles) long. More than 112 million tons of limestone were removed from this quarry during its period of use.

The finest Roman marble quarry was located at Carrara and operated beginning about 100 BC. It was often said that Augustus Caesar found Rome a city of brick and left it a city of marble. Roman taste in building admired size; the 80-ton blocks of marble that stand in Trajan's column were dwarfed by a block cut for the Temple of Bacchus at Baalbek, Lebanon, also in the second century AD, which measures 21 by 4.3 by 4.3 meters (68 by 14 by 14 feet) and weighs 1,500 tons. A great weight and volume of substance removed from the earth was required for the production of concrete and mortar, building materials that were first used to a great extent by the Romans, perhaps because of their strength combined with relative inexpensiveness. Mortar was known as early as Minoan times in Greece. Concrete was used in bridges and harbor works, which used enormous volumes of volcanic material from around the Mediterranean basin and beyond. Salt mines were plentiful in Mediterranean lands, but most salt was made by evaporation from seawater, or from lakes such as the Dead Sea and Lake Tatta in Asia Minor.

Mining and quarrying had pervasive impacts on the ancient landscape. Herodotus said that a whole mountain on the island of Thasos had been turned upside down and left in ruins by gold miners. A forest was removed by the excavation and by felling for fuel and timber. Wertime (1983, 448) writes that the mines at Laurium inflicted "a great scar upon the Attic landscape," and "by the time of Strabo the wooded surface of the region had been completely bared to provide timber for the mines and charcoal for the smelting of the ore." The ancient quarries of Giza, Syracuse, and Mount Pentelicus are still visible. Mining also diverted enormous quantities of water, much of it near the headwaters in the mountains, which deprived farmers lower down of a large portion of the supply they depended on and polluted what was left. Air quality was another concern in mines. Contamination came from gases trapped underground, and from the fumes of fires used for lighting the tunnels and for breaking rocks. Conditions in the workplace environment were appalling.

Metallurgical industries processed ores to recover the useful metals. They used tall furnaces that were often provided with chimneys. Writing of metallurgists in Spain, Strabo observes, "They build their silver-smelting furnaces with high chimneys, so that the gas from the ore may be carried high into the air, for it is heavy and deadly" (*Geography* 3.2.8). Other smelters were excavated as pits in the ground. Smelting required large amounts of fuel to reach the desired high temperatures. Technology advanced enough in Roman times to permit reworking of slag from earlier operations. Metalsmithing required more fuel and produced additional pollution. Minting of coins generated demand for precious metals. The mint of Rome consumed 50 metric tons of silver a year during the first century BC; many coins were then 94 percent pure. Each ton of silver required removing about 100,000 tons of rock from the mines.

Pottery was one of the most prolific industries throughout the ancient world. Ceramic factories required prodigious quantities of fuel to heat the kilns. Egyptian and Roman glassmakers did a lot of business, creating additional demands for fuel. Kilning of limestone for fertilizer, plaster, and mortar got its material from quarries, but in times of war and social upheaval might also use fragments of buildings and statues. To supply one limekiln for one burn in the Greek mountains required a thousand donkey loads of juniper wood, and fifty kilns required annually 6,000 metric tons of wood.

Along with household use in cooking and heating, industry produced a demand for fuelwood and charcoal that contributed to widespread deforestation. In the Mediterranean basin, at least 70–90 million tons of slag from the Greco-Roman period are known, the equivalent of 20–30 million hectares (50–70 million acres) of trees. Of course the deforestation was not quite as catastrophic as those figures might suggest, since the actual use was spread over nine or ten centuries. To smelt 1 ton of silver, however, required 10,000 tons of wood. It has been estimated that one operation, the Roman iron-smelting center at Populonia, used as much wood annually as is produced by 1 million acres of Mediterranean forest. The centers of mining and smelting became the areas most depleted of forests; copper mining in Cyprus and Western Sinai was especially destructive. In the latter area, now devoid of trees, archaeologists have found huge deposits of wood ashes (Thirgood 1981, 57). Faced with these great demands on sylvan resources, Rome increasingly turned to northern Europe, where extensive forests still survived. Even the Gauls complained of wood shortage. Coal was known, but it was considered an emergency fuel to be used only where no wood was available. It was burnt in Campania to make bronze because of the wood shortage there; elsewhere it was shipped by canal to the treeless marshlands.

Air pollution came not just from wood and charcoal smoke but also from the fumes of hazardous substances that were heated or burnt. Lead, often a major component of silver ore, was known to be dangerous. Vitruvius listed the symptoms suffered by lead workers, including complexions afflicted by pallor and loss of "the virtues of the blood" (*On Architecture* 8.6.10). Analysis of ice cores from glaciers in Greenland shows a marked increase in lead concentration during the second century BC, just when the Roman smelting increased sharply.

Vitruvius also worried about lead in the water supply. Lead pipes or aqueducts whose joints were sealed with lead could have contaminated acidic water, although most flowing freshwater in the Mediterranean area is in contact with limestone and therefore charged with calcium carbonate, a material that counteracts acidity. Calcium carbonate can also be deposited as travertine inside aqueduct channels and pipes, thereby isolating the water from lead. The Greeks and Romans were in greater danger of lead in their food, particularly acidic food prepared and served in lead and silver vessels. The effects of lead poisoning include infertility and impaired nerve activity in the brain and elsewhere. Tests of bones from ancient burials have shown elevated levels of lead. Mercury poisoning may also have presented a problem, at least for workers who smelted the metal from cinnabar, or who worked in gold, where mercury was used to make amalgam in a process that vaporized some of the mercury.

EPIDEMIC DISEASE

An important aspect of human interrelationship with the environment through history is communicable disease. Many human practices made its spread possible. With their population crowded together, the early cities offered an environment conducive to epidemics; this is undoubtedly one reason why life expectancy was lower among urban dwellers than in hunting and gathering tribes. Ancient armies, living in close quarters and coming into contact with other peoples, often suffered more deaths from pestilence than from hostile action.

In an age when bacteria and viruses were unknown, epidemic diseases were thought to be the work of the gods. As William McNeill notes (1976, 71), gods of pestilence are mentioned in Mesopotamian and Egyptian texts. The Hebrew Bible states that God sent a series of plagues, one or two of which may have been epidemics, to punish the Egyptian pharaoh when he would not release the Israelites from slavery. It further states that Philistine and Assyrian armies were stricken during and after attacks on Israel and Judah, and that David's people themselves lost thousands to disease after he conducted a census that God had forbidden. The *Iliad* proffers that Apollo sent a pestilence upon Agamemnon's army in retribution for an insult to his priest. But the gods could also avert illness. The god Pan used dreams to reveal a remedy for plague to the magistrates of Troezen. The Romans imported Asclepius, god of healing, from Greece to avert a pestilence.

Human populations are involved in a coevolutionary race with disease organisms. Each attempts to develop resistance to the defenses of the other. Without methods of inoculation, ancient human populations developed immunities to communicable diseases only at great cost in human life. But disease organisms mutated, producing new genetic strains that might overcome immunities and produce varying symptoms. Because diseases thus change through time, it is not always possible to decide exactly which diseases caused ancient epidemics. For example, the plague that enfeebled Athens in the Peloponnesian

War is difficult to identify. Although Thucydides gives a clinical description of its symptoms, they do not exactly match those of any modern disease. Hippocrates, however, does describe cases that sound remarkably like bubonic plague, so it seems probable that that well-known disease appeared in the Mediterranean area by the fifth century BC.

Epidemics clearly affected the course of ancient history. The Carthaginian siege of Syracuse in 396 BC ended when a plague decimated the attacking army. Had they succeeded, Rome might have come closer to losing the First Punic War. The Roman republic suffered many epidemics; Livy mentions a dozen or more, including one in 461 BC that killed both cattle and people, and adds that corpses too numerous for burial were thrown into the Tiber River. Pandemics seem to have increased in frequency when Alexander's expedition, followed by Roman trade, made contact with South Asian populations, bringing microbes home to add to native Mediterranean pathogens. After Augustus, Rome experienced plagues of increasing severity—not surprisingly, since merchants then journeyed regularly to India and even reached the court of the Han emperors in China. A pestilence under Nero killed thirty thousand. Even worse was the plague during the reign of Marcus Aurelius, the symptoms of which were described by the medical writer Galen. It was brought to Rome in AD 164 by soldiers returning from Mesopotamia, and it killed as much as one-third of the population; two thousand deaths a day in the city of Rome were reported at its height. Plagues were disasters for Rome's enemies, too; incursions of Huns and Vandals were blunted by them. Bubonic plague swept the known world in 540–565 under Justinian. It entered the Mediterranean through Egypt from a source in India or China, and is said to have halved the population of the late Roman Empire.

The effects of epidemics on Mediterranean civilization were significant. Human populations usually rebound after attacks of pathogens because survivors tend to be resistant, and birthrates rise as if to replace lost numbers. But if wars and other diseases interfere, losses may be repaired more slowly. Plague is associated with famine and declining agricultural production, since farmers may die from the disease or flee from the districts it attacks.

Malaria differs from the plagues described above in that it is chronic, endemic to local areas where its mosquito vectors reside, and may produce debilitation over a long period rather than sudden death. Skeletal evidence suggests that it was present in the eastern Mediterranean basin even earlier than Neolithic times. It was present in Egypt from at least as early as the First Dynasty (3200 BC), as immunological studies of mummies from that period have demonstrated. An inscription in the temple of Dendera warned those who read it to stay indoors in the hours after sunset while the Nile flood was receding (good advice, since the pools it left were perfect breeding places for mosquitoes, although the connection between mosquitoes and malaria was unknown at the time). Authorities differ as to whether it was present in Greece before the fifth century BC, and recent literature presents good reasons to think that it arrived either in Neolithic times or by the eighth century BC (Sallares 2002), but it was definitely known there in the fifth century and had become a prevalent disease by the fourth century. Reports credit the philosopher Empedocles (492–432 BC) with a project to deliver the Greek colonial city of Selinus in Sicily from fever by draining a marsh. Although he did not know that mosquitoes carry malaria, this would have been an effective way to deal with them. Malaria was noted in the swampy lowlands in and near Rome in the fourth century BC, and Romans devoted themselves to worshiping a fever goddess in the hope that she would protect them against it. There were at least three temples to the Dea Febris in the city of Rome. At the same time, ironically, every house and villa of the affluent class in Rome and its environs had an atrium with an impluvium, a pool to catch rainwater, which was a perfect breeding place for mosquitoes (Sallares 2002, 95). Poorer people living in the upper stories of *insulae* (apartment houses) without such an amenity were undoubtedly safer. Malaria took its name from the "bad air" of wetlands. Varro came near the truth about its vector when he advised anyone establishing a farm that "precautions must . . . be taken in the neighborhood of swamps . . . because there are bred certain minute creatures which cannot be seen by the eyes, which float in the air and enter the body through the mouth and the nose and there cause serious diseases" (*De Re Rustica* 1.12.2). From this vague passage, it is hard to know whether to credit him with making hesitant steps toward the germ theory or toward the realization that mosquitoes carry malaria. Columella added, equally imprecisely, that marshland "breeds insects armed with annoying stings, which attack us in dense swarms," and other things "from which are often contracted mysterious diseases" (*On Agriculture* 1.5.6). Malaria's effects worsened many problems affecting ancient Mediterranean societies. Farmers abandoned rich alluvial lowlands that might otherwise have produced good crops, and many of them moved to cities, where they swelled the numbers of propertyless urban poor. The affluent, of course, had been the first to depart. The result was waning agricultural productivity, not just because land was abandoned but also because those who had contracted malaria but not yet died from it were debilitated and, discovering that exertion brought on renewed attacks, avoided work. Malaria is not only something that impinges on humans from the environment, however. Its range can be extended by human modification of the landscape by actions such as deforestation, with the flooding, erosion, siltation, and creation of marshlands that result.

CONSERVATION: SACRED GROVES AND PARADISES

Ancient societies designated certain areas and preserved them in a relatively natural state. Some were sacred groves kept as places to worship the gods, and others were parks used for various forms of recreation by monarchs and citizens. In either case, sections of forest natural or planted survived, and wildlife found refuge.

The people of early times regarded particular groves as especially sacred, and these were protected from damage and everyday use. Sacred groves or forests dotted cities and the surrounding landscape as well. Often these contained temples. Public or private authority surveyed their boundaries and often marked them with stone walls. Sacredness meant that the trees within them could not be felled and their branches could not be cut. Even dead leaves and wood had to be left where they fell. Birds and other animals were safe from hunting within them, and the fish in their springs, lakes, and streams were guarded from fisherfolk. They were an important institution for which we have evidence from inscriptions and written leases, in addition to passages by many ancient writers.

Such a grove and everything in it were sacred. One would not lightly undertake the cutting of a tree there, and hunting was forbidden. Sacred groves existed in ancient Sumeria, Egypt, Babylonia, Assyria, and Persia as well as Greece and Rome. The practice of worshiping in special forest localities, and protecting them from destructive uses, is found in early human societies virtually everywhere that trees existed. It is known ethnographically among hunter-gatherer groups. Among farmers it is more noticeable because they cleared the land around the groves. Along with trees, the sacred areas usually contained water in the form of springs, streams, rivers, or lakes. The groves thus protected retained much of their original forested character and served as refuges while neighboring lands were transformed. As societies became urban, the building of temples and the planting of fruit trees or ornamental trees, possibly including some that were not native to the area, altered many groves. In other cases, builders planted new groves around the temples. Thus some groves did not retain the primeval qualities that once distinguished sacred forests; they became more like parks and gardens. The association with gods, spirits, and rituals, however, remained.

The Sumerians set their temples within groves and gardens. For example, in the city of Lagash, a garden called "the shade-of-the-plain" surrounded the temple of Ningirsu. A clay tablet that bears a plan of the city of Nippur, circa 1500 BC—one of the oldest known maps—indicates a *kirishauru*, a "park in the center of the city." This was not a palace park, Samuel Kramer explains, since *kirû* means a special garden, orchard, or palm grove connected with temples (Kramer 1959, 233). Temples serve to demonstrate the sacred character of these parks, but many sacred groves existed before temples; their trees and holy spaces were, indeed, the first temples, preceding the buildings that later usurped their name.

Egypt had temple gardens; indeed, an Egyptian temple had to have an enclosed grove and a sacred lake to be complete. Queen Hatshepsut brought incense trees from Punt to plant in the garden of the temple of Karnak, and her nephew Thutmose III followed her example by returning from his conquests in the north with countless trees and exotic plants, as well as birds and other animals. Ramses III gave 514 gardens to various temples. In Egypt's arid environment, they provided welcome shade and greenery. Historians of religious architecture have noted that temple interiors, with their forests of columns supporting the ceilings and clerestories, resemble groves of trees, and the dark hypostyle halls of Egyptian temples are good examples of this. Evidence of "paradises" in Egypt comes from papyri of Ptolemaic and Roman times. Virtually every village had these sacred groves, probably survivals from pharaonic times.

There is also evidence of temple gardens and sacred groves in Akkadian Babylonia in the time of Hammurabi, and among the Elamites and Kassites. Assyrian art shows altars in groves of trees. Assyrians held their New Year's festival with sacrifices of bulls and wildlife in a sacred park. Such a park was said to be an image of Mount Lebanon. Queen Semiramis planted a paradise at Behistun, a mountain sacred to the weather god, and used a large spring to water it.

"Paradise" derives from the Persian word *pairidaeza*, meaning an area of land enclosed by a wall. The Persian elite often used paradises for recreation, including hunting, but the original purpose of these reserved parklike sections of landscape was as sacred enclosures. The early Persian religion was polytheistic, and although Zoroastrianism, as the rulers' official monotheistic faith, replaced it, there were survivals of older beliefs and symbols, as in Mithraism, a mystery cult that flourished into Roman times. Tree symbolism was prominent in Mithraism; according to one myth, Mithras was born from a tree. Mithraic icons often show two trees, the trees of life and death. Similarly, there were two trees in the biblical Garden of Eden, the tree of life and the fatal tree of the knowledge of good and evil.

Zoroastrianism retained the traditional Iranian respect for trees. Zoroaster taught that trees are good creatures of Ahura Mazda, the supreme deity, and that tending them is an act of reverence. Modern Zoroastrians make pilgrimages to sacred groves, groups of huge old trees growing near springs, waterfalls, and streams. They honor certain trees by tying ribbons on them after making vows and offerings. The latter custom is widespread among many religions in the Near East.

The Greeks and Romans dedicated and protected hundreds of sacred groves. They varied in size: some were only a few trees around a temple that must have looked from a distance like the cypresses that fill walled cemeteries next to churches in modern Mediterranean lands. Plato's Academy was in a sacred grove dedicated to the Athenian hero Academus. An entire island, Icarus, was dedicated to the goddess Artemis, the guardian of wild animals. The grove of Daphne was ten miles in circumference, the sacred land of Crisa covered many square kilometers, and a grove near Lerna stretched all the way down a mountainside to the sea. Pausanias says that mountains bordered the sacred grove of the healing god Asclepius at Epidaurus on every side. Judging from the lay of the land there, it must have been extensive.

In Roman times there were countless sacred groves. Since they numbered in the hundreds, the area included in them was considerable, perhaps as much as 10 percent of the total forest land, and the practices used within them must have helped to conserve the biological diversity for which they served as refugia. The emperor Julian (ruled AD 361–363), who attempted to stem the Christian tide that was inundating the Roman Empire, sacrificed to the old gods in a paradise near his palace. There came to be Christian paradises, too, and they resembled sacred groves. The Christian emperor Justinian (ruled AD 527–565) dedicated a shrine to the Virgin in a place called a paradise, with a grove of cypresses, a meadow, and a spring.

To move from groves dedicated to gods to parks owned by kings was not to abandon the idea of the sacred according to the ancient mind, since sanctity surrounded the sovereign. Recent excavations in a palace at Amarna, the city of residence for the monotheistic Egyptian pharaoh Akhenaton, have revealed a walled garden containing a pool; a room adjoining the garden had mural paintings of lush vegetation and birds of many species. There were Assyrian royal parks stocked with many wild animals and with a variety of tree species; texts speak of these as symbols of the king's power and fertility.

The Babylonians, after the downfall of Assyria in 612 BC, had many royal parks, including the Hanging Gardens that Nebuchadnezzar built to please his wife, Amyhia, a Median who longed for the forested mountains of her homeland and asked the king to imitate them through the artifice of a garden. This was a pleasure garden, although the resemblance of the structure to a ziggurat, a kind of step pyramid that bore a temple at its summit, is unmistakable.

The Persian king had his own hanging gardens in Susa, one of his capitals. The Persian monarchs enjoyed their paradises, enclosed plantations of trees stocked with animals and birds that they sometimes hunted. They were well maintained and sturdily walled to prevent entry of unauthorized persons and to protect them from damage. Their natural beauty was unspoiled. Stately forest trees grew in them around water features. There also could be formal or informal gardens in them. When paradises contained fruit trees or spice trees, as they often did, the products could be gathered, perhaps for use in the palaces or for sale. Although the sources rarely mention using trees in paradises for wood, the biblical book of Nehemiah records that King Artaxerxes I sent a letter to Asaph, guardian of the royal paradise, directing him to provide timber for Nehemiah's rebuilding of the temple and city of Jerusalem.

Affluent landowners in Greece and Italy imitated royalty with private retreats. Persian paradises had impressed visiting Greeks, including Xenophon, who saw many of them, introduced the word into Greek, and designed his estate according to the patterns he had observed in Persia. Further to the west, Dionysius the Elder, ruler of Syracuse, planted plane trees in a paradise in Rhegium. Alexander the Great not only visited many Persian paradises but as conqueror of the Persians, and therefore their king, became owner of them. He had a throne erected for him in one, where he sat when conducting royal business. When he had his final illness in Babylon, his friends took him on a boat across the river to bathe and rest in a paradise there.

CONCLUSION

The landscapes of the Mediterranean basin and Near East have suffered greatly from human occupation since ancient times. They have been subject to cycles of devastation and recovery, and much of the devastation now visible is the result of medieval and modern mistreatment of the natural world. But ancient peoples did degrade their environments, even if not to the extent of modern times.

The fact that many of the problems of human ecology as they are now understood also existed in ancient times should not come as a surprise. These include deforestation and overgrazing that removed vegetative cover; erosion of the land; destruction of wildlife; pollution of air, land, and water; depletion of resources; agricultural decline; and manifold urban difficulties such as food and water supply and sewage disposal. The city cannot be discussed in isolation from the countryside that it impacts and upon which it depends, for both are part of the same ecosystem.

The relationship of ancient societies to the natural environment was determined in part by their characteristic mental constructions of nature. The actions of people reflected their perceptions and values, even if many of these were in conflict with one another. Ancient attitudes varied from worship of nature to curiosity, desire for prudent use, and greed leading to wasteful exploita-

Bare hills behind the small port on the island of Hydra, Greece, show a deforested state all too typical of the Mediterranean islands. Most of the vessels in the foreground are fishing boats, although Hydra also has a merchant marine school. (Photo courtesy of J. Donald Hughes)

tion. People were seldom deliberately destructive, and they often ameliorated the condition of the world in which they lived by planting parks and gardens and by protecting certain areas such as sacred groves. At the same time, natural and economic forces could distort and overwhelm reason, custom, and religion. People often are not aware of the long-term results of their actions upon the natural world. Intending to sustain balance, they nevertheless upset it. As John McNeill remarks, "Nothing alters ecology quite like civilization" (McNeill 1992, 71). If ancient people adversely affected the natural environment within which they lived—and they certainly did—it is reasonable to suspect that they may have helped to bring about the decline of their own civilizations. The end of the ancient world therefore had an ecological dimension.

The sight of ruined cities surrounded by ruined land has long been a source of comment by travelers in the region. All around the Mediterranean, ancient ports have been landlocked by erosional sediments, a fact mentioned by ancient writers. In the former Roman provinces of North Africa, the wide avenues and impressive buildings of cities such as Lepcis Magna, Sabratha, and Thamugadi, which once exported wheat and olive oil, stand empty within a Saharan landscape that could not support such populations today without the importation of large amounts of water. The irrigation systems of the Carthaginians and Romans, which once made this arid region blossom, depended for their effectiveness on watersheds whose forests have disappeared. Although it would be incorrect to blame the ancient inhabitants for all the defects of the present-day Mediterranean lands, since they have been subjected to successive pressures, it seems clear that the ancient peoples in many instances initiated a process of wearing away the environment that had supported them. The environmental changes they caused brought pressures on their societies and created problems for those who followed in later periods.

References

Diamond, Jared. 1997. *Guns, Germs, and Steel: The Fates of Human Societies.* New York: Norton.

Greig, J. R. A., and J. Turner. 1974. "Some Pollen Diagrams from Greece and Their Archeological Significance." *Journal of Archaeological Science* 1: 177–194.

Jennison, G. 1937. *Animals for Show and Pleasure in Ancient Rome.* Manchester: Manchester University Press.

Kramer, Samuel Noah. 1959. *History Begins at Sumer.* Garden City, NY: Doubleday.

McNeill, John R. 1992. *The Mountains of the Mediterranean World: An Environmental History.* Cambridge: Cambridge University Press.

McNeill, William H. 1976. *Plagues and Peoples.* Garden City, NY: Doubleday.

Meiggs, Russell. 1982. *Trees and Timber in the Ancient World.* Oxford: Clarendon Press.

Pritchard, James B., ed. 1958. *The Ancient Near East: An Anthology of Texts and Pictures.* Princeton: Princeton University Press.

Pyne, Stephen J. 1997. "Eternal Flame: Fire in Mediterranean Europe." In *Vestal Fire: An Environmental History, Told through Fire, of Europe and Europe's Encounter with the World,* 81–146. Seattle: University of Washington Press.

Rice, E. E. 1983. *The Grand Procession of Ptolemy Philadelphus.* Oxford: Oxford University Press.

Sallares, Robert. 2002. *Malaria and Rome: A History of Malaria in Ancient Italy.* Oxford: Oxford University Press.

Sandars, N. K., ed. 1972. *The Epic of Gilgamesh: An English Version with an Introduction.* London: Penguin Books.

Thirgood, J. V. 1981. *Man and the Mediterranean Forest: A History of Resource Depletion.* London: Academic Press.

Tringham, Ruth. 1971. *Hunters, Fishers, and Farmers of Eastern Europe, 6000–3000 BC.* London: Hutchinson and Co.

Wertime, Theodore A. 1983. "The Furnace versus the Goat: The Pyrotechnologic Industries and Mediterranean Deforestation in Antiquity." *Journal of Field Archaeology* 10, 4: 445–452.

3

THE MEDITERRANEAN
IN THE MIDDLE AGES

I am setting on the Earth a steward.
—QUR'AN 2

Praised be you, O Lord our God, king of the universe, who has not left a thing lacking in his world and created in it good creatures and good trees, so that human beings can benefit from them.
—JEWISH BLESSING

Man has nothing more than the beast. Of earth they were made, and into earth they return together. What then is man but slime and ashes?
—INNOCENT III, *ON THE MISERY OF THE HUMAN CONDITION*

Be praised, my Lord, through our sister, Mother Earth, who sustains and guides us, and brings forth diverse fruits along with colorful flowers and herbs.
—FRANCIS OF ASSISI, *THE SONG OF BROTHER SUN*

When the Middle Ages began, the eastern Mediterranean basin was in the hands of the Eastern Roman Empire (also known as the Byzantine Empire), a polity that preserved Roman law, spoke the Greek language, and observed the Christian religion in its eastern forms. Its memorable emperor Justinian I (527–565) had reconquered for the empire the lands it had lost to the Ostrogoths in Italy and the Vandals in North Africa, and even part of Visigothic Spain, but these gains were subsequently lost. The Byzantines and Sassanid Persians contested for Mesopotamia and the Levant, but soon both those empires were fighting for their lives against a new power that arose in an unexpected quarter: the Arabian Peninsula.

The ascendancy of Islam, and of the Arab armies whose success carried it forward, altered the map of the Mediterranean basin and brought about far-reaching economic and environmental changes. By the time of the death of the prophet Muhammad in 632, Muslim rule embraced most of Arabia, and within twenty years after that the Muslims had taken Egypt, Syria, Palestine, and Persia (Iran). In a second wave of conquests under the Umayyad caliphs, they swept across Libya, Tunisia, Algeria, and Morocco, and over the Strait of Gibraltar to take Spain in 711. An invasion of France failed at the Battle of Tours in 732. The Abbasid caliphs made Baghdad their capital in 750, and presided over an Arabic flowering of literature, art, science, and medicine, stimulated by the study of the Greek classics. Many classical writings were preserved in Baghdad, but the last survivals of the culture that found its center there were almost snuffed out by the Mongols, who in 1258 looted the city, throwing the contents of the libraries into the Tigris River to build a "bridge of books" that stained its waters dark with ink. The Mongols, however, being a continental people from the landlocked regions of Central Asia, found the waters of the Mediterranean Sea to the west of their newly conquered region to be an alien environmental realm, and they did not acquire a navy or try to continue by sea the conquests they had so overwhelmingly achieved with their archer cavalry on the Eurasian landmass. This failure of adaptation proved incredibly fortunate for the western Mediterranean coastlands, both to the north and to the south of the inland sea. Christian kingdoms gradually pushed Muslim rule out of Spain during the Middle Ages, capturing Granada, the last Muslim stronghold, in 1492.

The Byzantine Empire managed to hold Asia Minor against the Arabs but lost its holdings in Sicily and southern Italy to western commanders. The capital, Constantinople, was seized by the Fourth Crusade in 1204 at the instigation of the city of Venice, which had provided two hundred ships for the expedition, and was regained by the Byzantine Greeks under Michael VIII Palaiologos only in 1261. The city and the empire itself succumbed to the invasion of the Turks, originally a Central Asiatic people, at the end of the Middle Ages in 1453. Turkish forts subsequently stood on both sides of the Bosporus and controlled trade between the Mediterranean and the resource-rich lands to which the ports of the Black Sea were the keys.

By the latter part of the period, swathes of forest on the northern margins of the Mediterranean basin were being removed, and new machines were being used; plowing took place faster over longer stretches of field, and trade revived and extended farther. European and Arab traders voyaged around the Mediterranean Sea. After the depredations of the Black Death, there were again more people around the inner sea. Built-up areas were spreading, and with them, forest clearance, erosion, and advancing desert. Parts of the surrounding continents, if less than before, were still covered with forests, and if some places might have looked like wilderness, in fact peoples had lived there for centuries or millennia. The rate at which humans were altering the face of the Earth was slow but accelerating. It was not proceeding at a steady pace, but it was clearly faster than it had ever been before. Certain societies of the Mediterranean were acquiring skills that would in future times become more effective. They were learning to learn about the world—haltingly, with insufficient methods—but learning nonetheless. In the age to come, some of them, first of all the Portuguese and Spaniards, would break forth upon the rest of the Earth. Preparations for rapid modern changes in the world environment were made in the Middle Ages.

HUMAN SETTLEMENTS

During the Middle Ages, well over nine-tenths of all people in the Mediterranean area lived on the land in a rural setting. Even before the Roman Empire had entered its eclipse, estates belonging to wealthy patrons were moving toward autarky all around the Mediterranean. They were selling less over long distances and depending more on the local manufactures of the manors. The items they sought in trade were not raw materials to any great extent, but luxuries such as spices, precious woods, and metals, and a few necessities such as salt. Honey, wax, and furs came from the north shore of the Black Sea. Trade in many of these commodities was conducted by the Byzantines from ports including their magnificent harbor at Constantinople and to a lesser extent by other easterners including the Jews.

In the eastern Mediterranean area, new Arabic cities flourished but were situated away from the sea, compared with the ancient centers of the same regions. In Egypt, Greek-speaking Alexandria on the Mediterranean coast was replaced as capital by the Arab foundation of Cairo 190 kilometers (120 miles) up the Nile River. In Syria, the Umayyad rulers preferred Damascus, screened from the sea by the double wall of the Lebanon and Anti-Lebanon mountain ranges, to the old Seleucid capital of Antioch, only 20 kilometers (12 miles) up the Orontes River from its mouth. Tripoli and Algiers were exceptions. The Abbasids far inland in Baghdad contentedly left trade on the Mediterranean and Black seas to Constantinople. This was not due to a lack of Arab interest in navigation, since Arabs were magnificent sailors who were to bring the compass from China, but perhaps a preference of the Abassids of Baghdad to direct their own trade toward the procurement of luxuries by way of the Persian Gulf, Arabian Sea, and Indian Ocean. The pattern characteristic of the Islamic cities was that of an organic

The commercial power of the city of Florence, Italy, in the thirteenth century, which had widespread effects on the European environment, is represented by the city hall (Palazzo Vecchio) and the cathedral of Santa Maria del Fiore, viewed from the Uffizi Palace. (Photo courtesy of J. Donald Hughes)

arrangement of inward-facing buildings around courtyards with few open spaces, quite different from the "Hippodamian" square-block plan of most of their Greco-Roman predecessors. The Islamic plan was adapted to the provision of separate neighborhoods for communities based on kinship, tribe, and ethnicity including the recognized non-Muslim groups. Fez in Morocco is a good example of this plan. In the Byzantine world, Constantinople retained its unrivaled supremacy even during the period of Venetian rule, but urban growth of centers away from the capital was relatively minor.

The western Mediterranean was something of a backwater in the early Middle Ages. The French historian Henri Pirenne suggested that this was because the Arabic conquest left the Mediterranean split into a Muslim south and Christian north, with resultant barriers to trade. Some cities, Marseille for example, still existed within their Roman walls and retained the Roman gridiron street plan. Seacoast towns in southern Europe moved to hilltop locations partly to avoid Arab pirates, but also to escape the "bad air" of coastal wetlands that caused, people thought, malaria. Of course the real vector of malaria was the mosquitoes that bred in the very same wetlands. But the reason for the decline of trade was more probably the effects of the barbarian invasions in the west and the localization of the economy. In any case, commodities from the eastern Mediterranean continued to find their way to the west through Syrian Muslim and Jewish intermediaries, and tin, textiles, and other western Mediterranean products went under sail to the east. The character of the use of natural resources and their dissemination through trade had greatly altered by the eleventh century with the growth of urban centers in northern Italy. Port cities such as Venice, Genoa, and Pisa increased rapidly in population and engaged in trade across the Mediterranean Sea. Florence, Siena, and Milan, although located inland, also became centers importing natural resources and engaging in manufacturing. Florence eventually conquered Pisa and became a sea power in its own right, with galleys that sailed to east and west.

The Crusades undoubtedly caused a sea change in the relationship between the western and eastern halves of the Mediterranean. Originally intended as military campaigns of the Christian West blessed by the pope with the object of recovering holy sites from the Muslims, these expeditions became the instruments of the new western trade centers in their own economic interests. Unfortunately a number of diseases, including influenza, were inadvertently spread between the eastern and western Mediterranean lands by the crusaders. Although ultimately unsuccessful in reconquering the Holy Land, the Crusades left Venice and Genoa in control of important trading bases in the eastern Mediterranean. Western ships were capable of carrying cargoes of five hundred tons by the thirteenth century, and the trade between French and Italian ports and the Levant in importing to the western countries products such as spices, perfumes, textiles like raw cotton and silk, dyes, and sugar had become quite lucrative.

Trade and financial dealings also increased in the opposite direction between the cities of northern Italy and the growing urban centers of northwestern Europe. A major mover of economic expansion was the wool trade; it was an important activity of great companies such as the Bardi and Peruzzi, both Florentine firms. It supplied the necessary resource for an industry that produced fine woolens from imported raw wool. Giovanni Villani in the 1330s stated that there were two hundred shops belonging to the wool guild in Florence, employing thirty thousand workers (Villani 1955, 71–74). They turned out eighty thousand bolts of cloth annually, selling them for 1.2 million florins. A large portion of the water supply from the Arno River's clear tributaries was used to wash wool and provide energy for water mills used in fulling cloth. Cloth makers took every bit of wool the local sheep produced, and looked to distant sources. Merchants drew on fleeces from England, southern Italy, North Africa, and Spain.

Many people today look back on the medieval period as a time of stagnation between Classical civilization and the vigorous onset of science and industry in the modern age. Nothing could be further from the truth. Population almost tripled in the northern sections of the Mediterranean basin during this period, and the number of settlements increased proportionately. In fact, such rapid expansion of economy and population over so great an area—around 2 percent per

Rome's River Tiber, a source of water, an avenue for boat traffic, and a sink for sewage disposal, illustrates the complex relationships of human society to the natural environment. (Photo courtesy of J. Donald Hughes)

year—had not occurred before in history. Although this is not extreme by modern standards, it was then unprecedented over such a long period. This momentum of population growth stretched the limits set by the environment under the conditions of medieval knowledge and technology. Towns grew into full-fledged cities and began to face the problems of waste disposal, pollution, water supply, flooding, and even air pollution. Northern Italian cities began to deal with the crises of urban sanitation through law codes and improved infrastructure.

Water supply was a problem for cities throughout the Middle Ages. With the collapse of Roman imperial administration, the aqueducts continued to function only if some local authority such as a city government or bishop took responsibility, and because the aqueducts were often very long and the damage expensive to repair, this seldom happened. This meant that city folk depended on wells, cisterns, and water carriers, and that water quality became undependable due to inadequate supplies and pollution. By the eleventh century, town officials in northern Italy had begun to supply piped water, and similar provisions were slowly adopted elsewhere. Sewage was a terrible danger to health in the medieval Mediterranean due to its contamination of the water supply. Wells were often in low-lying places, and as Zupko and Laures remark (1996, 63),

"The streets, which were drained by open gutters, received all sorts of refuse and excreta and tended to drain into these same low-lying areas." The same authors quote a law of 1276 in Verona that indicates the desire of some city governments in the Middle Ages to maintain a safe water supply by cleaning a spring called Masera:

> Since the water of the Masera is quite useful and well known, and since many persons, both native-born and foreigners, come and go to these waters for the sake of their bodily health, we decree and ordain that, on behalf of the honor and advantage of the Commune of Verona and of the persons coming to the aforesaid Masera, the Masera should be cleansed, repaired, and renovated for the Commune of Verona, to the extent that it seems expedient. (Zupko and Laures 1996, 66)

Just how expedient the city fathers might have considered it in actual practice is questionable, since they also passed a law prohibiting pollution of the Adige River during the day but permitted its use as a sewer at night. Urban sanitation seems to have improved as the Middle Ages continued, but the whole period was "low-lying" compared with the sanitation provisions of the Roman Empire in former times.

TECHNOLOGY

Technological inventions prepared the way for modern attempts to control nature, but they also enabled management of the environment to a significant extent during the Middle Ages. Windmills, foreshadowed by earlier experiments in Alexandria and subsequently adapted to practical use, supplemented human and animal energy in tasks such as raising water for irrigation, the grinding of grain, the sawing of wood, and the cutting of stone. Water mills had been used widely since Roman times for the same purposes, and their use for those activities continued and expanded. Improved sailing vessels made exploration and long-distance trade more possible than before, along with introductions of exotic species and products valuable enough to be worth carrying. Chinese inventions such as iron plows, clocks, magnetic compasses, padded horse collars, and cannon passed to the West through Arab intermediaries or a few western traders and adventurers such as Marco Polo.

In the early Middle Ages there was a deterioration of technology in the portions of the Mediterranean basin that had been subject to barbarian invasions. Some valuable skills had been lost, so that the quality of products such as cloth, pottery, and metallic ware was coarser. In contrast, the Byzantine Empire had preserved and even improved Roman technology, and the Arab world demonstrated an increasing mastery of techniques. But by 1200 the initiative in technology had been seized by the West, to a large degree as a result of adopting the achievements of Arabic science and innovations that had been passed to the West by the Arabs from civilizations further to the east, such as the numerical system including the zero, originally invented in India. Three inventions of the high Middle Ages made possible the fine wool textile industry mentioned earlier: the spinning wheel, the horizontal pedal loom for weaving, and the fulling mill powered by water.

In the fifteenth century, humanistic writers directed some of their very active curiosity toward practical technology. Leone Battista Alberti (1404–1472), the Venetian polymath whose many artistic and architectural achievements included designing the Trevi Fountain for Pope Nicholas V, authored a treatise on shipbuilding entitled *Navis*. In it he examined the remains of Roman ships found submerged in Lake Nemi in Italy in order to discover aspects of their functional operation that might prove useful to designers in the Arsenal, the shipyard in Venice.

By the end of the Middle Ages, the blast furnace had revolutionized the production of iron, which greatly increased as a result. Lynn White, a twentieth-century historian of medieval technology, has argued that Western medieval Christianity, as a monotheistic religion that uniquely gave human beings dominion over the created world, produced an intellectual atmosphere conducive to the advances of science and technology. That may be too simple an explanation, however, since Eastern Orthodoxy, Islam, and Judaism also taught similar ideas of stewardship and dominion, and some of the most important discoveries were made in the East.

AGRICULTURE AND PASTORALISM

Agriculture was the basic economic activity in the Middle Ages, and both wealth and power were based on landholding. That is, they ultimately rested upon the peasantry—the vast majority of the population—and the peasants in turn subsisted upon the land and the forms of life that inhabited it, domestic and wild. The fate of human societies and the fate of the landscape were every bit as intimately linked as they had been in ancient times.

Production still emphasized the ancient Mediterranean triad of grains, olives, and grapevines. One new crop, rice, was introduced by the Arabs from the east and found its most congenial place in terrain that was naturally flooded

Monasteries had their own estates and in some Christian lands became centers of agricultural recovery, producing cereal grains, grapes, olives, hay, and animal products. In Muslim southern Spain, the Moorish nobles introduced cotton and other crops, restored Roman aqueducts and other works, and, building more of their own, made possible intensive irrigation agriculture. A similar system was used in Syria. Egypt, as one might expect, remained a unique case where the annual flood of the Nile River provided a rich and dependable harvest of bread grains almost every year, from which beer was also made. Egypt also produced wine, particularly in the Delta region, and it appears that in spite of a reduced supply due to the Qur'an's prohibition of alcoholic drinks, the output neither of wine nor beer was entirely intended for export or limited to the substantial Christian and Jewish minorities. New crops such as sugar cane, citrus fruits, melons, and strawberries came into Islamic agriculture, were adopted in Spain, and eventually spread to other parts of the northern basin.

Cotton, a fiber that had been raised and used in India since ancient times, was only a minor competitor with wool in the Middle Ages. Greek travelers, who called it the "wool tree," had brought samples from there before the time of Alexander the Great, and it was planted in Roman times as a garden curiosity, but it was only after the Muslim conquests that it became an item of trade. Cotton cloth was first imported from India, and then Egypt and other Middle Eastern countries—and eventually the Muslims in Spain—began to raise cotton and export cotton and cotton textiles across the Mediterranean. In the late Middle Ages cotton became popular in Italy and was often mixed with wool. Eventually the Sicilians obtained seeds and began their own cotton industry. Cotton farmers were forced to contend with the problem of soil exhaustion; they had to use various kinds of fertilizers, and Moorish cultivators in Spain alternated cotton with plantings of alfalfa (its name is Arabic), which they observed to improve the fertility of the soil.

In the European section of the Mediterranean basin, agriculture entered a period of rapid growth from about AD 1050 to 1300. Many landholders encouraged their peasants to open lands to farming that had been woodland, marsh, or moor, and as a result new villages also grew up. The great Florentine historian Niccolò Machiavelli opined that the extension of agriculture into new areas was advantageous to princes and republics alike. Those who led society had to be concerned that food production be able to keep up with the needs of an increasing population, and one of the most effective ways to do this in the Middle Ages was to bring new land under cultivation. More farmers indeed meant more production, but this could only remain true until every acre of arable land was utilized. If that occurred, where could new farms be opened? Giovanni Boccaccio described the building of new terraces on the mountainsides to extend culti-

A farm on the shore of Lake Lugano, a lake in the Alps shared by Italy and Switzerland, reflects the timeless nature of agriculture in the rural Mediterranean. (Photo courtesy of J. Donald Hughes

or prone to flooding. This meant that lands formerly too marshy for crops became arable, notably in the lowlands behind Venice. Elsewhere in the Mediterranean area the crops and the methods of cultivation had changed little from Roman times. The heavy moldboard plow drawn by four to eight oxen (reported in the first century AD by Pliny the Elder as an innovation in southern Germany) and the three-field system, increasingly important in northern Europe, had limited application, even for wheat, to the southern limestone-based soils and dry Mediterranean climate, where the two-field system continued in use. They were used to some extent in the Po Valley in northern Italy. In the early Middle Ages, farms and farmers suffered greatly from wars and invasions, and in some regions peasants fled from their lands to the relative safety of towns. Production dropped, even of such staples as olive oil and wine, and the population initially decreased.

Southern Europe did not adopt feudalism to the extent of the north, although great manor houses reemerged in southern France and adjacent northern Italy. The Byzantine Empire had a feudalism of its own, and inner Asia Minor, still somewhat protected, was a reservoir of population and a source of troops.

vation, but terracing required great amounts of work to build and maintain, and the bulk of the remaining land might be so marginal that working it might not be worth the effort either financially or in terms of labor.

Before 1300, food supply mostly remained adequate, and though failures in distribution produced local shortages, no widespread famines are recorded. In the early fourteenth century, with little new land available for agricultural expansion, the increase in production failed and the manors resorted to desperate moves. Serious famine occurred every ten years or so. Larger cities found that neighboring agricultural lands no longer offered adequate staple crops. Florence may serve as an example of this trend. A Florentine grain merchant, Domenicho Lenzi, reported in the early fourteenth century that the surrounding territory in Tuscany produced only enough grain to feed the city for five months of the year. The rest had to be imported, mostly from southern Italy, and its import was one of the most important activities of the large trading companies. Weather, crop failures, and war made supply insecure. Thus it is not surprising that many cities tried to supply at least part of their food needs by gardens and plantations within their urban centers. Churches in Constantinople, for example, encouraged the planting of the Mediterranean triad (grains, olives, and grapevines) in the open land that they owned inside the city, including the land right around the church buildings, a project that provided them with extra income.

Although by this time horses were a source of energy for plowing, many others were used for war, and all ate quantities of oats that might have fed the increasing numbers of hungry peasants. Fewer oaks, due to deforestation, meant less pork, since acorns were a major food of pigs. The medieval village was a sustainable ecosystem when it had the expansive landscape of earlier times to interact with, but in the overcrowded fourteenth century it proved unstable.

Crop failures fell upon Italy from the 1320s onward. Florence suffered along with other cities; food prices were higher than ever. Famine struck in 1329, and the price of wheat rose three to five times above former levels. Starvation returned in the fateful year of 1339, a time when it was difficult to pay for imports because the city had a huge war debt. This was the time of the *condottieri*, unscrupulous leaders of bands of mercenary soldiers that roamed the countryside and offered "protection" to cities that would hire them, and they demanded to be fed, too.

Grazing animals, including cattle, sheep, goats, and pigs, were important sources of food, clothing, and other products in the Middle Ages, as they had been in ancient times. In addition, horses, mules, and donkeys provided transportation and drafting. These animals could graze on the fallow fields, enriching them with manure, and also on marginal lands. Transhumance, the practice of driving the herds up into the hills during the hot summer when vegetation at

Sheep grazing on tender spring grass near Nauplion, Greece. Sheep are notoriously destructive of vegetative cover, and excessive grazing, especially in the highlands, contributed to soil erosion. (Photo courtesy of J. Donald Hughes)

lower elevations was dry and unpalatable, continued as in earlier centuries. Shepherds and cowherds watching over the flocks had indeed been a feature of the Mediterranean landscape for millennia, with both good and bad results for the natural environment. In addition to producing manure, the grazers reduced the amount of low vegetation, lowering fire danger somewhat. But they often set fires to encourage regrowth and improve the grazing conditions. Not everyone agreed with this practice; a Spanish monarch, appropriately named Peter the Cruel, ordered that a person caught starting a fire should be thrown into it. Overgrazing resulted in barren hillsides that allowed both flooding and erosion that wasted good soil.

As a source of wool, the most important raw material for the textile industry, sheep were the largest portion among the numerous grazing animals, even allowing for their smaller size. The demand for wool caused a rapid augmentation of the flocks, increasing the impact on forests and grasslands. Sheep are notoriously destructive of vegetative cover, and excessive grazing, especially in the highlands, contributed to soil depletion. New breeds of sheep, such as the merino of Spain, bore more and better wool, but they also tended to strip the soil of vegetation more efficiently. In Spain, transhumance took place over long distances between the warm south in the winter and the cooler north in the summer. This activity came under the control of a cartel of sheep owners called the Mesta, which appeared in the late thirteenth century, lasting until the nineteenth, and gained control over the trails, or *cañadas*, on which they moved their flocks twice a year, with the right to use them unhindered by stoppage on the part of peasants or landowners. One of the less happy results of intense grazing was that while greater numbers of people were being clothed, the land was being denuded.

FORESTRY AND DEFORESTATION

In the early medieval centuries from about AD 600 to 800, after the decline of Rome and with the disruption of economic activities caused by the barbarian invasions, the forests of the northern margins of the Mediterranean had a chance to reestablish themselves to some extent through natural reproduction. The violence and depopulation of the period had at least one unintended but positive result—the regrowth of the woods. Between 800 and 1000, there was incipient economic recovery, a growth in the population, and increased use of forest products. This trend was greatly accelerated in a new period of rapid population growth and revival of the economy between about AD 1000 and 1300. Population had doubled between 600 and 1000, and doubled again between 1000 and 1200. The clearing of forests was one of the most visible changes in the landscape. Other changes, such as the expansion of agriculture, the exploitation of minerals including the smelting of iron and other metals, the development of water power and wind power, and the revival of transportation networks, involved demands on the supply of wood and directly or indirectly caused forest removal.

The arrival of Islam on the southern and eastern Mediterranean coastlands and in Spain meant that the early medieval period was an age of conquest, military and economic activity, and trade, not a "dark age" at all. Timber was in great demand, especially for shipbuilding as the Arabs turned to the sea, and for other forms of construction. Cairo and Tunis became great shipbuilding centers. Because wood was in short supply along the eastern stretches of the Mediterranean's southern shores, Egyptian shipbuilders (many of these were Coptic Christians working for Muslim contracts) had to commission wood from northern Syria and the Maghreb, or gratefully buy it from timber pirates who raided the Anatolian coasts of the Byzantine Empire. Forests also had to provide fuel for metallurgy, ceramic manufacture, and sugar refining. Since the older centers of civilization such as Syria, Mesopotamia, and Egypt—now under Arab, or at least Muslim, rule—had already to a large extent exhausted their sylvan resources (as,

The Taurus Mountains are seen here above the countryside of southern Turkey. Most of the land around the Mediterranean Sea is mountainous. (Wolfgang Kaehler/Corbis)

for example, the cedars of Lebanon), it was necessary to look to the west and north for timber of construction quality. With the conquest of the Maghreb, the Atlas Mountains became an advantageous source of wood, since they were then still forest-covered and had their own cedars among other fine species. Other partly forested areas that were in the Muslim sphere for a time included Iberia, southern Italy, and the Mediterranean islands of Sicily, Crete, and Cyprus. The Taurus Mountains on the southern margin of Asia Minor provided an excellent supply of tall, straight trees, and although they were for centuries still in the hands of the Byzantines, the Arabs were well equipped and located to trade or raid along that coast.

The period of rapid population growth in western and central Europe, including Italy and France, from about 1050 to 1300 saw a transformation of the landscape to one where forests had been reduced to isolated fragments. Settlers saw the woods as a barrier, and not only cut them down but sometimes even deliberately burned them off. The forest gave way before the axe, saw, and the farmer's plow. A major purpose of forest removal was to expand the area under cultivation and pasture, thus increasing the wealth of the nobles and churchmen who controlled the land.

But far from being worthless wastes, forests actually represented an indispensable source of resources for the economy. Wood and charcoal, its partially oxidized product, were virtually the only fuels for heating, metallurgy, and manufacture of glass, tile, bricks, and pottery. Using wood in the form of charcoal requires consuming five times more material by volume to produce the same amount of heat energy than if it is burned directly. Wooden torches supplied lighting. Carts and ships, weapons and musical instruments, dishes, and shoes were commonly of wood. Wines were stored in oaken barrels. Castles and fortifications required timber. Even stone buildings had wooden roofs and required scaffolding during construction. All these uses hastened forest removal.

A large swath of the Mediterranean lands suffered deforestation even greater than what had occurred in ancient times. In the eleventh century the mountains of northern Italy had good forests, many of which had regrown in the centuries since Roman exploitation. Large tracts of unexploited forest of oak and chestnut were granted to monasteries in that century. Collection of pig rent indicates a sylvan landscape, since pigs found acorns and other favorite foods in oak forests. Records indicate that during the following three centuries sheep and cows, which prefer grassland pasture, came to outnumber pigs in this area.

One factor that may have saved some areas from total deforestation was the practice of royalty in reserving some tracts of forest for the purpose of hunting. This hunting was, of course, reserved for the king and other nobles. The reservation of royal forests was most common in northern Europe, but some of the French reserves were in Mediterranean southern France. King Alphonso IX of Leon and other Spanish monarchs enjoyed hunting bears and wild boars, and they retained as royal property large tracts of forests for this function. The king of Portugal made certain areas game preserves (*coutadas reais*), mostly located near the central coast and the Tagus River valley where they would be accessible to the court, but one or two were created in the more isolated northern region. Forest wardens were appointed to safeguard these areas, although it is true that they also received certain hunting privileges as perquisites of the job, which may have lessened their protective effect. Pine trees were planted to stabilize coastal sand dunes at Leiria, Portugal, as early as 1325. Princess Eleonora of Sardinia proclaimed a law protecting nesting hawks on the cliffs of the island, which as a result came to be called Eleonora's falcons. Peasants found their customary activities restricted by various kinds of royal reservations, and poaching of animals, birds, and timber was not unknown even though it might be punished severely. Some species were reduced in number or extirpated by hunting. Elephants disappeared from their last North African stronghold in the eleventh century; this was due partly to hunting and the ivory trade but also to the continuing defor-estation of the Atlas Mountains and consequent desertification.

In the same period, the Italian coastal cities of Venice, Genoa, and Pisa (un-til its conquest by Florence) and others built formidable navies in order to de-velop and protect marine trade throughout the Mediterranean. By the year 1000, Venice had used its naval strength to clear the Adriatic Sea of pirates. The needs of shipbuilding in centers such as the Arsenal in Venice, which was at the time the greatest industrial center of Mediterranean Europe, placed severe demands on the forests for timber, masts, and forest products other than wood, including pitch for waterproofing. Venice also used immense numbers of trees as pilings driven into the marshy soil to support its houses and churches. Seeing the dan-ger that suitable timber might not remain in their own territory, thus causing them to need to expand their already extensive purchases of timber from abroad, the doges of Venice enacted conservation laws in the thirteenth and fourteenth centuries, limiting the quality of wood that could be burnt as fuel in the glassmaking industry to that unsuited for ships and setting a precedent in preserving forest resources. The doges also prohibited unauthorized export of timber from the districts of the neighboring Alps, which they were able to dom-inate. Genoa in Italy and Barcelona in Spain, with well-forested mountains at their backs, also became important shipbuilding centers.

Other cities also tried to safeguard wood supply. The statutes of one com-mune near Siena in 1281 required every member inheriting a portion of land to plant ten trees a year. Measures such as these met with at most partial success, and the pattern of forest removal continued. Dante set the beginning of his *Di-vine Comedy* in a dark forest in the year 1300, but there was little forest re-maining near his home city of Florence then. Stone and marble often replaced scarce timber in building. Shortages of charcoal for metallurgy appeared, and bricks, which require firing and therefore fuel, became more expensive. Wine prices increased due to a scarcity of oak for casks; in fact, "by the end of the thirteenth century the price of wine was determined by the availability of casks rather than the quantity or quality of the vintage" (Bowlus 1980, 89). Loss of tree cover increased the number and severity of floods as water from storms poured down denuded slopes. Florence, located on the banks of the capricious Arno, was and is vulnerable to flooding, as in the case of the disastrous flood of 1333, which broke all four of the bridges and inundated the city center. Erosion of slopes due to deforestation led to declining soil fertility in the highlands and exacerbated the return of sluggish malarial swamps to a lowland coastal zone that also suffered from encroaching salinity. To give another example, in the thirteenth century Louis IX founded the port of Aigues-Mortes on the French Mediterranean coast in the western part of the Rhône delta and sailed from there when he undertook the Seventh and Eighth Crusades. His son Philip III completed fortifications of the town and used it in his ill-fated campaign against Spain. In the course of the following century, however, the silting up of the coast had begun to make the port unusable. It is ironic that the cutting of trees for the construction of naval vessels for these kings further up in the watershed of the Rhône River may have had a role in shortening the useful life of the very port they constructed for their navies. The shrinking of forests was a pivotal cause of the environmental crisis of the fourteenth century.

EXTRACTIVE INDUSTRIES

Mining and metallurgy improved in several regions, providing materials for tools used in agriculture, forestry, and hunting, and weapons for war; the mines and smelters themselves increased demand for fuelwood and charcoal, depleting forests and producing pollution.

The manufacture of one ton of iron required the annual increment of thirty acres of productive forest. Some metals were imported; gold came from West Africa by camel caravan across the Sahara through the ports of the Maghreb, and the Venetians tapped the trade in silver, lead, and copper from the interior of the Balkan Peninsula.

Iron was used early in the Middle Ages for armor and weapons, horseshoes, and tools as large as plowshares and as small as fishing hooks and needles. It also found its place in construction; doors needed hinges and locks. Furnaces of the time were inefficient hearths and ovens, located if possible in exposed locations where the wind would hopefully create a dependable draft and fan the fires. Later, hearths were provided with bellows operated by human or animal muscle or by waterwheels. These furnaces, which were extremely wasteful in their use of fuel and therefore tended to be located in remote forested areas where water power was also available, produced a bloom of soft iron that could be hammered to get remaining impurities out, and then reheated to be worked into the desired shapes. Monasteries sometimes exploited ore deposits, as did the Carthusian monks in northwest Italy. The eastern Alps and the Pyrenees along the French-Spanish frontier provided supplies of ore. Reworking of iron into fine tools and weapons such as swords became a specialty of guilds in towns such as Toledo in Spain, Milan in Italy, and Damascus in Syria. Later on, the scale of iron metallurgy increased, and better models of furnaces appeared, eventually leading to the invention of the blast furnace at the end of the medieval period. Coming to the Mediterranean from north of the Alps, the blast furnace was more efficient in its use of fuel and produced a purer iron, but consumed the resources at an increased

Sicilian men harvest sea salt from evaporation beds near the island of Mozzia. This has been a major economic use of Mediterranean Sea waters in ancient, medieval, and modern times. (Jonathan Blair/Corbis)

rate due to the growing scale of production. The problem was not exhaustion of the ores, since the scale was as yet too small for that, but the need for a continuing supply of charcoal, the required fuel. After using up the forests on the surrounding hills, a center of production might have to close.

Salt, of great value for the preservation of fish and other meats in these days before refrigeration, was made by evaporation in salt pans in western Sicily, southern France, the Adriatic near Venice, Illyria, and elsewhere along the seacoast, as it is still done today, and its sale to inland districts helped to make coastal towns prosperous. The activity took up wetlands that were nesting territory for waterbirds and rich breeding grounds for fish.

RELIGIOUS ATTITUDES ABOUT THE ENVIRONMENT

In the Mediterranean region during the Middle Ages, the dominant religion in the European north and the Byzantine east was Christianity in its two major forms, eastern and western; Islam dominated the African south and Middle Eastern areas; and Judaism was an important presence and influence everywhere. All three of these great monotheistic religions had ambivalent attitudes toward nature. They shared the ideas that the natural world is the creation of God, and that humankind is part of that creation with a special role delegated by God. Just what that role was, however, varied within and among the three religions. There were also conflicting ideas about the attitudes that humans should have toward the natural world and its living denizens. Most monotheists saw God as transcendent, beyond and above the natural world, and many of them also thought that attribution of sacredness or high value to created things was at least questionable and perhaps heretical.

In Christianity, a prevalent attitude about creation was expressed by the powerful thirteenth-century pope Innocent III, who saw man as equal to the beasts, but held that this equality lowered man. In his *De miseria humane conditionis*, he wrote:

> The Lord God formed man from the slime of the earth, an element having less dignity than others. . . . Thus a man, looking at sea life, will find himself low; looking upon the creatures of the air, he will know he is lower; and looking upon the creatures of fire he will see he is lowest of all . . . for he finds himself on a level with the beasts and knows he is like them. Therefore the death of man and beast is the same, and the condition of them both is equal, and man has nothing more than the beast. Of earth they were made, and into earth they return together. What then is man but slime and ashes? (Innocent III, *On the Misery of the Human Condition* 1969, 6–7)

The pope thus urged Christians to denigrate the created world and turn their attention to heaven and the salvation of their souls. Practical expression of this attitude had been provided long before, when St. Benedict had cut down a pagan sacred grove to prepare the land for his monastery at Monte Cassino.

The attitude that the natural world is inferior to that of the spirit, and therefore is of relatively minor significance, persisted even in those who in the late Middle Ages began to turn to the study of the Greek and Latin classics and to embrace a new humanism that presaged the Renaissance. Petrarch (Francesco Petrarca) was one of the most noted of these figures. In 1336, at the age of thirty-two, he decided to climb Mont Ventoux in the French Alps above his home in Avignon. Though at only a little over 1,900 meters (6,200 feet) in elevation the peak is far from daunting, to undertake such a feat was quite rare in those days and has been taken to indicate an interest in wild nature. As Petrarch described himself, "What am I? A scholar? No, hardly that; a lover of woodlands, a solitary, in the habit of uttering disjointed words in the shadow of beech trees" (Petrarch 1966, 45–51). His motives for the climb were curiosity and the desire to emulate Philip V of Macedonia, who had ascended Mount Haemus in Thessaly because he had heard that one could see both the Aegean Sea and the Euxine (Black) Sea from its summit. Petrarch was amazed at the view from the top of Ventoux, but he had happened to bring along a copy of St. Augustine's *Confessions*, and when he consulted it, it fell open at a passage that read in part, "Men wonder at the heights of the mountains, . . . but themselves they consider not." On the hike back down, Petrarch kept silent, his glance turned downwards rather than at the landscape, and on his return to the inn wrote his father, "How earnestly should we strive, not to stand on mountain-tops, but to trample beneath us those appetites which spring from earthly impulses" (Petrarch 1966, 45–51).

A contrasting Christian attitude to nature is embodied in a number of Christian mystics, including Hildegard of Bingen, Mechtild of Magdeburg, Meister Eckhart, and Julian of Norwich. Hildegard, who wrote music as well as poetry, said that the Word, the incarnate God, manifests itself in every creature. Notable among this group is Francis of Assisi, whose most prevalent image in Christian iconography shows him preaching to birds or shaking hands with the paw of a wolf. Both of these stories are connected with La Verna, a mountain that he climbed and where he spent forty days, recalling Jesus's forty days in the wilderness. Of course many saints long before had demonstrated their holiness by their association with animals—St. Jerome had his lion, St. Euphemia her bear, and St. Humbert his stag—but Francis went beyond them. Francis epitomized a way of appreciating and treating nature, not the dominant mode in Western Christian thought and action, but an attractive view and a possible corrective. He was deeply sympathetic to the powerless among humans and also among all living creatures. His central characteristic was his insistence on the goodness of creation. All his writings about created things are positive; he sees God in them, worships God through them, and gives thanks for them. Francis is original in using terms of family relationship to refer to created entities. He addressed the sun, moon, fire, water, plants, and animals as brothers and sisters, recognizing in them the same origin as himself. Earth is a special case, since he called it both sister and mother, and said that it governs humans and other creatures like the mother in a family. Francis further valued creatures for their individuality, and his behavior toward living things can be seen as an ethical application of reverence for each kind of creature in the diversity of creation. He emphasized the presence of God in the diversity of created beings and urged humans to rejoice in this diversity and glorify God for it and with it, and act in ways consistent with respect for it. If God's grace is mediated to people through elements such as water in baptism, wine and bread in Holy Communion, and oil in several of the sacraments, why cannot it also be received in an analogous way from any creature?

The Franciscan monastery of the Carcieri outside Assisi, Italy. Francis of Assisi is said to have spent time here when the area was a wild forest. The area immediately around the monastery is protected and has a dense, high forest; the hillsides above have been damaged by grazing and woodcutting. (Photo courtesy of J. Donald Hughes)

Certainly an innovation of Francis was his practice of preaching directly to nonhuman creatures. A sermon to the birds is the most widely known incident of his life, and he also preached to fishes and flowers. Francis took literally the scriptural command to preach the gospel to all creatures. In doing so, he recognized their intrinsic worth along with humankind as fellow creatures of God. He extended the concept of community beyond the religious order to every human being, and beyond the human race to every natural creature. The story of the wolf of Gubbio illustrates this. The people of the town asked the saint's aid in stopping the depredations of the ferocious creature. Francis did not simply order the wolf to stop attacking animals and humans around the town; he established a covenant between the wolf and the human community, binding the people to feed the creature in return for the wolf's forbearance. The Franciscan covenant had other ramifications: he not only urged people to feed wild birds and animals in winter but also thought the emperor should make the practice mandatory, as well as providing feed for oxen and asses on Christmas Eve. He urged that laws be enacted to prohibit the killing of "our sisters the larks." The Franciscans, and other orders as well, tried to make the gardens of their monasteries little representations of the original Garden of Eden. Francis may be a medieval precursor of environmentalism.

Many Christian monks, including some of the followers of Francis, devoted much of their time to labor, in the garden and in the fields, that would transform the landscape. They felt that such work was of spiritual value: *laborare est orare* (to work is to pray). Humankind, they believed, remembering the words of Genesis, was put into the earthly garden "to till and to keep it," and the occupations of cultivation, arboriculture, and pastoralism made them partners of God in completing and improving the creation. The Cistercian monks deliberately located their monasteries in wilderness tracts so that they could direct their efforts to making them useful to human beings, and to glorify God. Such a positive view of human efforts to improve the Earth was not limited to monks, of course. The Duc de Berry, brother of the king of France, commissioned a famous *Book of Hours* in 1409, with paintings by Pol Malouel de Limbourg and Jean Colombe, that portrays the round of agricultural labors and other outdoor pastimes through the year in a ducal demesne, including woodcutting in February, plowing the fields and trimming the grapevines in March, fishing with nets in April, all the way to beating down acorns for the hogs in November and hunting boar in December. All of these activities received the obvious approval of the artist, and in the upper parts of the paintings the stars revolve to bestow the evident blessing of God.

Islam teaches that God created the Earth for the use of humankind, and that principles for that use are contained in the Qur'an and in Islamic tradition. God forbids environmental disruption: "You must not exceed the measure, but observe the measure strictly, and do not fall short of it. And the Earth He has appointed for His creatures" (Qur'an 55:8–10). Creatures are part of a community; the Qur'an recognizes something like the ecosystem: "There is no species of animals on Earth, nor birds on the wing, who are not part of a community, just as you are" (Qur'an 6:38). Man is the representative or viceroy of God on Earth, and the authorities of the Islamic state are instruments of God. The Islamic state, therefore, must control people's actions so as not to allow the environment to deteriorate. Humans are to be kept aware that this world is only a temporary abode, and that they must work in this world in such a way as to be rewarded in the next world, their permanent place of residence. In terms of natural resources and environmental quality, Islam teaches that exploitation that harms another cannot be permitted. If a person neglects the land, therefore, the community embodied in the Islamic state may take it away and ensure its proper treatment. The Shari'ah (Islamic law) contains a number of positive environmental provisions. For example, it declares an inviolate zone around wells to prevent their pollution. Establishing a principle of humane treatment for living beings, it decrees that any animal killed or slaughtered must be killed as painlessly and quickly as possible. Practices ascribed to Muhammad himself include the establishment of public reserves for conservation and production, including the protection of wildlife and the grazing of horses for the Muslim army, and the prohibition of the cutting of any tree in the desert that provides shade or food for humans or animals. He declared that anyone who restores desert lands should acquire title to them. The caliph Abu Bakr prohibited devastation of an enemy's crops, orchards, and livestock, allowing his army to take only what was necessary for food, a command that he based on statements of the prophet Muhammad. Islamic jurists permitted hunting for food, but not for sport or trophies. Muslims believe that if Islamic law were enforced equitably and obeyed, there would be proper use of resources and environmental damage would be minimized or would disappear.

Along with this legalistic view, and somewhat in contrast with it, is another Islamic tradition that was widely represented in the Middle Ages, that of the Sufis. The Sufis embraced the unity of God as pervading all nature, and they sought oneness with it. The Iranian poet Sa'di wrote ecstatically that he was in love with the whole universe because it comes from God. The Sufis emphasized the immanence of God without denying the transcendence. Their identification with God in nature sees the environment as a whole, in which everything is interrelated, not just on the physical plane but also spiritually. In the Sufic view, humans can love nature and contemplate its never-ending forms as theophanies of the Divine All-Possibility. There is a Divine Book, the recorded Qur'an, and there is also a Qur'an of Creation that contains the archetypes of all things. Many Christian mystics, possibly including Francis and Paul, felt the same way about the relationship between the Holy Scriptures and the invisible nature of God that can be seen in the Creation.

The Jews were a minority throughout the Mediterranean in the Middle Ages, but their scriptures were older than those of Christianity and Islam and contained ideas about the relationship of humankind to God and nature that can be seen to have influenced the other two religions. Judaism affirms that the world belongs to God, its creator. This is reason for praise and thanksgiving. A psalm addresses God, who "set the earth on its foundations, . . . makes springs gush forth in the valleys, . . . and causes the grass to grow for the cattle. . . . O Lord, how manifold are your works! In wisdom you have made them all; the Earth is full of your creatures" (Psalm 104:5, 10, 14, 24). A traditional Jewish blessing thanks God for leaving nothing lacking in the world and for creating

The Dome of the Rock, a mosque in Jerusalem, marks the site of the prophet Muhammad's heavenly journey. This beautiful example of Islamic architecture reminiscent of the natural environment, the dome of the sky, is near to places also hallowed by Jewish and Christian monotheistic traditions. (Photo courtesy of J. Donald Hughes)

beautiful creatures and beautiful trees for which mankind may glorify God. Furthermore, much Jewish literature affirms that the Creator can be seen in creation. The psalm says, "The heavens declare the glory of God; the firmament proclaims his handiwork. Day to day pours forth speech, and night to night declares knowledge" (Psalm 19:1–2). In the book of Job, Job himself says,

> *Ask the beasts, and they will teach you;*
> *The birds of the air, and they will tell you;*
> *Or speak to the Earth, and it will teach you;*
> *And the fish of the sea will declare to you.*
> *Who among these does not know*
> *That the hand of the Lord has made this?*
> *In God's hand is the life of every living thing*
> *And the breath of all mankind. (Job 12:7–10)*

Later in the same book, God's voice out of the whirlwind presents a majestic picture of creation, one of the finest evocative passages ever written, as unchallengeable evidence of the power of God.

In the view of the Jewish scriptures, human beings are not the lords of creation, free to do with the Earth whatever they please, but God's stewards, responsible to God for their actions. The basic principle of the treatment of nature that derives from this vision is the command *bal tashhit* (do not destroy). The Torah (the law embodied in the first five books of the Bible) contains many environmental commandments, of which a few examples follow: It forbids even an army besieging an enemy city from cutting down trees (Deuteronomy 20:19). A mother bird may not be taken with her young (Deuteronomy 22:6–7). Cities should have an open space a thousand cubits (1,500 feet) wide around them, free of construction and cultivation. Later Jewish tradition continued these concerns. God is portrayed speaking to Adam in Eden, telling him not to corrupt or desolate the world, since there will be no one other than humankind to set things right later on. A species may not be destroyed; the Talmud has the raven rebuke Noah, who is about to send him out of the ark, where there are only two ravens, "If sun or rain overwhelm me, would not the world be lacking a species?" (Sanhedrin 108b).

Jewish scriptures present the picture of an agricultural society, but in the Middle Ages Jews became increasingly urbanized, perhaps either because their lands were seized or because of the advantages and comparative freedom offered by city life. It was also a time in which poetry and mysticism increasingly flourished. The great scholar Maimonides wrote that a person could learn to love God through the contemplation of God's created works. The Kabbalistic writings expanded on this idea, seeing nature as a garment of God. Like Francis's poem quoted above, "Perek Shira," a mystical poem from around 900, has verses from many different creatures singing praises to God. The mysticism of Abraham Abulafia included meditation in the natural environment. The mystics of Safed developed seders for Tu B'shvat, a kind of Jewish "Earth Day" or "Arbor Day" including the planting of trees, to celebrate the presence of God in nature. Thus they and many others did not simply contemplate but advocated kindness to creatures and ethical behavior toward the rest of creation.

The degradation of the natural environment, however, was not prevented by monotheistic religious traditions either in the Middle Ages or in other periods. This is in part due to the failure of the traditions to adapt to the situations presented by changing ecological conditions. Then too, religious perspectives and prescriptions do not exclusively determine human behavior. It is an unfortunate fact observed often in history that people, in their quest for material benefits, have disregarded the ethics of their own religions. Commandments have often proved to be inadequate motivation for human beings, especially when those commandments seem to contravene economic self-interest.

In the high Middle Ages, especially in the European part of the Mediterranean, technology had its own pattern of development seemingly independent of religion. It improved the processes of mining and metallurgy. Waterwheels, windmills, and sails employed the energies of nature for human purposes. But deforestation and local pollution were among the side effects of these seeming successes. Food production and distribution could not keep up with the needs of a briskly growing population, thus creating a seedbed in terms of crowding and debilitation for the imminent spread of an opportunistic organism that was the cause of a pandemic.

THE BLACK DEATH

The Black Death, or bubonic plague, arrived in the Mediterranean area in 1347. Beginning, some scholars of medical history believe, in the province of Yunnan in China, it had traveled by way of the Silk Road westward, taking advantage of the movements of the Khanate of the Golden Horde, reaching the shore of the Black Sea at the trading post of Kaffa (Theodosia) in the Crimean peninsula. From there, Genoese ships carried it to Constantinople, Cairo, and Messina in Sicily. Once established in a few ports, the disease spread on ships to virtually every port in the Mediterranean, and thence inland. Its rapid dissemination on shipboard is due, of course, to the fact that its vector is rat fleas (this was unknown at the time) and that black rats infested every vessel. Some cities adopted measures to control the spread—the word *quarantine* derives from a

Venetian practice—but in vain, especially since the mode of transmission was not understood. Between 1347 and 1351, plague killed at least one-quarter of the population of the lands bordering the Mediterranean. Florence, where it arrived in April of 1348, suffered greater than average losses, which were described graphically by Giovanni Boccaccio in the preface to the *Decameron*. Three-fifths of the Florentines—that is, about sixty thousand—died. Seven more outbreaks occurred around the Mediterranean in the following eighty years. Half the population of Sardinia had perished in the first outbreak, and half the survivors died in a second onslaught in 1376. The Mediterranean region was in economic and environmental crisis already. Agricultural productivity had declined due to the mistreatment of the land during the period of unrestrained population growth. The weakened condition of the people due to famine and lack of resources made the loss of life worse than it would otherwise have been. With fewer workers, labor costs and the prices of commodities rose. As French poet Guillaume de Machaut put it:

> *Thus it happened that for lack of people*
> *Many a splendid farm was left untilled,*
> *No one plowed the fields*
> *Bound the cereals and took in the grapes,*
> *Some gave triple salary*
> *But not for one denier was twenty [enough]*
> *Since so many were dead.*
> *(Herlihy 1997, 41)*

Some writers have suggested that the Black Death relieved the ecological crisis, reducing the population to a level that no longer pressed so hard on the carrying capacity of the land. Since there were fewer peasants to clear the land and to cultivate the soil, the forests inched back into abandoned fields and reoccupied some of the marginal lands that had been imprudently developed. These new forests would be ready for another wave of exploitation later on. The human population of the Mediterranean basin recovered, although it took a long time. As late as the mid-nineteenth century, Tuscany, as an example, would not have as many people as it did in 1300, about two million.

CONCLUSION

In the Middle Ages, the peoples of the Mediterranean borrowed environmental capital from their ecosystems and just as surely squandered it. They might have liked to renege on their debts, but unlike money debts, environmental debts cannot be renounced. In the fourteenth century, nature sent bill collectors in the shape of resource scarcities, famine, and perhaps even the Black Death. The population and economy had come up against environmental limits. The evidence shows that the medieval economy, at the level of technology then available, grew to such an extent that the ecosystems were no longer able to support it. Some of the results included famines and lowered resistance to epidemic diseases. Although farmers or incipient industrialists had no evil intentions in this regard, it was human activities that caused the crisis. The rapidly increasing population of the high Middle Ages, its growing economic activity, and the resultant depletion of resources had a negative effect on the relationship between humans and the environment.

References

Bowlus, Charles R. 1980. "Ecological Crises in Fourteenth-Century Europe." In *Historical Ecology: Essays on Environment and Social Change*, ed. Lester J. Bilsky, 86–99. Port Washington, NY: Kennikat Press.

Herlihy, David. 1997. *The Black Death and the Transformation of the West.* Cambridge, MA: Harvard University Press.

Innocent III (Lothario dei Segni). 1969. *On the Misery of the Human Condition.* Ed. Donald R. Howard. Indianapolis: Bobbs-Merrill.

Petrarch, *Letters on Familiar Matters* 4.1. Petrarch (Francesco Petrarca). 1966. *Letters from Petrarch*, translated by Morris Bishop. Bloomington: Indiana University Press.

Villani, Giovanni. 1955. *Cronica.* Book 12, chap. 94, quoted in *Medieval Trade in the Mediterranean World: Illustrative Documents Translated with Introductions and Notes*, eds. Robert S. Lopez and Irving W. Raymond, 71–74. New York: Columbia University Press.

Zupko, Ronald Edward, and Robert Anthony Laures. 1996. *Straws in the Wind: Medieval Urban Law in Northern Italy.* Boulder, CO: Westview Press.

4

THE EARLY MODERN PERIOD
(1492–1799)

> The Mediterranean soil too is responsible for the poverty it inflicts
> on its peoples . . . The thin layers of topsoil . . . are enabled to
> survive only by man's constant effort. Given these conditions, if the
> peasants' vigilance should be distracted during long periods of
> unrest, not only the peasantry but also the productive soil will be
> destroyed. . . . In the Mediterranean the soil dies if it is not protected
> by crops: the desert lies in wait for arable land and never lets go.
>
> —FERNAND BRAUDEL, *THE MEDITERRANEAN*
> *AND THE MEDITERRANEAN WORLD*, 1972

The early modern period saw the beginning of the age of European expansion and domination of the world economy, and it was two Mediterranean nations, Portugal and Spain, that initially led that expansion. Voyages under Spanish sponsorship found the way westward to the Americas, while Portuguese navigators rounded Africa, pioneered the route to India and the Spice Islands, and in the process opened the Atlantic islands and Brazil to colonization. The environmental effects of this worldwide spread of enterprise on the countries impacted by Europeans were overwhelming—and are beyond the focus of this book—but the reciprocal effects on the Mediterranean area were also great. The widespread adoption of American food plants is one example. The tomato, which is a native of South America, transformed Mediterranean cooking. Indeed, it is difficult to imagine what Italian cooking must have been like before the introduction of the tomato. Maize, domesticated in the New World and found by the Spaniards as a plant cultivated in the shadows of the pyramids of the Valley of Mexico, was soon to be planted in the shadows of other pyramids in the Valley of the Nile. Beside these beneficial plants, some troublesome weeds made the journey from the New World to the Old. The prickly pear, an American native, is something of an ambiguous case. It became ubiquitous in the Mediterranean area and served as an important food source as well as a planted fence, but it went wild and, armed with thorns that were injurious to humans and animals, made many fields virtually useless.

Other imports also modified the Mediterranean lands. A huge influx of gold enriched the rulers of Spain and Portugal, modified the economy, and inflated the price of land and other commodities. An upward spiral of production and consumption, interrupted but not ended by economic downturns, increased the pressure on natural resources.

The boundary between Christendom and the Islamic dominions tilted as the early modern period began. In the west, the Catholic monarchs of Castile and Aragon captured the last Moorish stronghold, Granada, in 1492, and from that time onward Spain controlled everything as far south as the Strait of Gibraltar. In addition, Portugal had a foothold in Africa at the port of Ceuta. The Mediterranean islands were also in western hands: the Balearics, Sardinia, and Sicily belonged to Aragon; Corsica to Genoa; and Crete and Cyprus to Venice. France, a great power of this period, held a section of the northern Mediterranean coastline and would add Corsica in 1768. But in the east, the Ottoman Empire followed up its capture of Constantinople by overrunning Greece and the Balkans as far north as the Danube River and beyond, and began an advance through North Africa that added Egypt and eventually Libya, Tunisia, and Algeria to its dominions. In 1529 the Turks besieged Vienna without capturing it, but also in the first half of the sixteenth century they succeeded in annexing Hungary, Bosnia, and Serbia. By 1571 they had taken Cyprus, and in 1669 Crete, from Venice. For most of the early modern period, therefore, the eastern Mediterranean and the Black Sea margins as well as almost the entire southern coastline were in Ottoman hands. In the eighteenth century the Turks were losing the northern Adriatic lands to Austria and Venice, and the northern Black Sea area to Russia. At the end of the century, the new order in the Mediterranean was indicated by the defeat of the French fleet at Alexandria in Egypt by a British fleet commanded by Admiral Horatio Nelson. The inland sea had become an arena of conflict between western European powers, a pattern that would persist until World War II.

HUMAN SETTLEMENTS

Most of the important Mediterranean cities retained their walls, gates, and towers throughout the early modern period in spite of the fact that cannons and gunpowder had rendered those defenses somewhat anachronistic. Within the walls, masonry, stone, and wood construction might be found. Streets were narrow, seldom paved, and almost without exception cluttered with offal. Horses provided transport but also deposited tons of manure on public ways. Sometimes street cleaners gathered the animal and human excreta and carted them out to the environs to be sold to farmers and gardeners for fertilizer. Public fountains provided water that was often polluted and therefore dangerous to health. Scattered outbreaks of plague continued, periodically reducing urban populations. For example, Verona had its population cut in half by an outbreak of plague in 1632 and did not fully recover until the early nineteenth century.

Northern Italy as a whole lost 16 percent of its population between 1600 and 1650, but this was not due to plague alone. As Fernand Braudel observes, "Plague . . . is only a passing visitor to the Mediterranean. Malaria is permanently installed there" (Braudel 1972, 64). The prevalence of malaria in low-lying places and its relative absence in the hills led some towns and villages to relocate to sites at higher elevations. The town of Ninfa, next to the Pontine Marshes, was abandoned between AD 1675 and 1680 because its location was judged to be excessively conducive to malaria. There were also attempts to drain swamps; Pope Sixtus V tried unsuccessfully to drain the Pontine Marshes between 1585 and 1590, and he died of malaria because he made the mistake of visiting his works in progress.

Plague and malaria are just two of the many diseases that were especially rampant in urban areas. Only constant immigration from the countryside managed to maintain and increase the size of cities, which even in times free from plague still experienced more deaths than births on the average. In Rome, for example, deaths exceeded births in every decade in the eighteenth century, and a considerable majority of the residents had been born outside the city. Food supply was always a problem; each Italian city customarily selected an official to be entrusted with the job of obtaining an adequate quantity of grain, and similar measures were sometimes used elsewhere. Malnutrition was a recurring problem. Many cities still included fields and gardens within their walls as a partial means of feeding the population during sieges or other times of emergency. These small open spaces were fertilized by city waste and usually irrigated from the city water supply.

Attempts to plan cities and provide them with wider avenues centered on public squares were rare. Some of the sixteenth-century popes tried to carry out such rebuilding in Rome, but their plans were left half-realized. Mediterranean cities had been built centuries earlier, and it was difficult to change what was there. Towns that dated back to Roman times sometimes possessed a center with the rectilinear plan of an original Roman military encampment, surrounded by a more organic (or chaotic) arrangement of streets that had grown outward during the Middle Ages. In other cases the population had decreased and there were unoccupied spaces inside the city, or ad hoc walls that had been thrown up, using stones and bricks from ruined houses to defend against barbarian raids, still surviving from troubled times.

Istanbul had reached a population of 300,000 by the 1640s, making it the largest city in the Mediterranean area (Venice was half as populous), and indeed the largest city in Europe. The supplies necessary for this great city came from throughout the empire and beyond. The closest lands had to provide everything that was within their capabilities. Sheep numbering in the thousands were driven from the European dominions near the city. Grains, vegetables, and fruits came by ship from seaports on the Aegean Sea and the Black Sea. The latter also provided meat, animal fat, leather, cotton, beeswax, salt, and slaves. Newly conquered lands, such as those on Crete, which fell into Ottoman hands in the late 1660s, were sold to private owners who the government believed would make them productive, and sometimes these were Jews or Christians, not always Muslims. The conquest of Egypt placed the ancient breadbasket of the Mediterranean in the hands of the Ottomans, and the trade with India came under their control.

A similar economic impoverishment of the hinterlands occurred in other large Mediterranean cities such as Madrid and Naples. Naples was the largest city in western Europe in 1595, with 280,000 people. Even so, it must be emphasized that the rural population made up the vast majority, at least 90 percent, of the inhabitants in the Mediterranean area, and that agriculture was by far the most important economic activity.

NEW SETTLEMENTS

A hitherto unknown section of the Mediterranean climatic zone, although located outside the Mediterranean Sea proper, came to the attention of the Iberian peoples and was occupied and significantly altered by them in the years leading up to and including the discovery of the Americas. The Canary Islands and Madeira have a Mediterranean climate and to a great extent a Mediterranean vegetation. They were known, but only slightly and for the most part by rumor, to the ancient Mediterranean peoples, including the Phoenicians, Greeks, and Romans, and called Macaronesia, a Greek word that means the "Fortunate Islands" or "Islands of the Blessed." When the Canaries were rediscovered by Europeans in the fourteenth century and settled by them in the fifteenth century, they were already inhabited by a people known as the Guanches, who had been there for centuries, farming and clearing land by burning, and keeping huge herds of goats that had devastated the landscape and caused severe erosion.

Columbus stopped by the Canaries on his way westward at a time when the Spaniards were still waging a campaign to take the islands from the Guanches, who had obtained some aid from Portuguese troops hoping to frustrate their Spanish rivals. The Spaniards, aided by a pestilence that they had inadvertently introduced to the Guanches, gained complete control of the islands by 1500, and within another generation it was hard to find more than a handful of the natives. The Guanches had already introduced Mediterranean species of animals and plants that did well in a climate that differed little from their places of origin. The European settlers continued this trend, bringing more domestic animals and crops, including sugar cane, which would become the basis of the export economy of Macaronesia until it was undercut by cheaper, more abundant sugar from Brazil.

Madeira was another story. Until a Portuguese settlement came in 1425, it had no human inhabitants, and was covered by a luxuriant forest of laurel and other native trees so thick and abundant as to give the island its name—*Madeira* means "wood" in Portuguese. It had many indigenous birds and insects, including a number of flightless species. The only mammals were bats and the colonies of monk seals—*lobos* (sea "wolves")—on the coast. Despite the shortage of wood in the Mediterranean heartlands, on Madeira the forest represented an embarrassment of riches; in fact, the settlers regarded it as an inconvenient barrier to agriculture and began to cut and burn it to clear the land. The folk memory on Madeira includes the tradition that, once ignited by the settlers, the forests burned steadily for seven years. Residents including Christopher Columbus, who had a house and lands on Madeira's smaller neighbor island of Porto Santo, remarked that the removal of the forests had caused a great reduction in the amount of clouds, mist, and rainfall on the island. Domestic animals munched the vegetation, and cats, mice, and rats (whether deliberately or accidentally introduced) wreaked havoc among the birds, making several species extinct. The first important crop was sugar cane, worked by slaves; after the collapse of the sugar market, the second important crop was vines to produce the famous Madeira wines, whose excellent flavor and keeping quality were discovered by accident when visiting ships carried barrels through the tropics. Exceptionally delicious bananas were also grown, but long shipping times and the small size of the fruit kept them from prospering in world markets. A settler released rabbits on Porto Santo, and their offspring soon became so numerous that they temporarily drove the human inhabitants away. This story of total transformation of the landscape, with the extinction of native species and the uncontrolled explosions of introduced populations, had undoubtedly been enacted on the islands in the Mediterranean Sea in past

North side of the Portuguese island of Madeira in the Atlantic. This island was first settled in 1425. A primeval Mediterranean forest, the Laurisylva, was removed and replaced by terraces on which sugar cane and the famous Madeira wine grapes are grown. (Photo courtesy of J. Donald Hughes)

millennia, and was to be repeated on countless other islands around the world when European ships made their visits. The story of the Canaries' Guanches, unfortunately, would also be repeated.

AGRICULTURE AND PASTORALISM

Sheep and goats had prospered in the Mediterranean lands since early times, enabling humans to produce food and clothing but at the same time eating up the vegetation, preventing the growth of trees, denuding hillsides, and creating the conditions for ruinous erosion. Grazing came to dominate the landscape over extensive sections of Spain, where the association of owners called the Mesta sponsored the clearing of forests to create pasture for burgeoning herds of Merino sheep, which numbered 3.5 million by 1526 and constituted the most profitable portion of the Spanish economy. The shepherds used trees for fodder and annually burned the vegetation to encourage the growth of grass, which in turn was subjected to the close cropping of sheep. The barren, eroded landscape that resulted astounded visitors from elsewhere, including Venetian ambassadors who described it in diplomatic correspondence, but similar processes of denuding were also at work in southern France and the Balkans, undoubtedly including some of the Venetian territories along the Adriatic seacoast and its fringe of islands. Spanish agriculture also suffered because the Moriscos, or Spanish Moors of Islamic origin, were expelled from the country at the beginning of the sixteenth century, and these were the very people with the most understanding and experience of irrigation agriculture. Agriculture fared no better in the Ottoman Empire, however, where any production more than subsistence that appeared was taxed away by the state to finance the military and the swollen bureaucracy of Istanbul. Still, Fernand Braudel believed that "the Mediterranean was able to live largely off its own agricultural produce" (1972, 423), a statement that was probably true in most years.

Agriculture in the Mediterranean basin during the early modern period retained the traditional crops, including the ancient triad of grains (wheat and barley), grapes, and olives, which were supplemented by legumes, green vegetables, and root crops. But new plants appeared from abroad, especially maize, the potato, and the tomato, all three introduced from the Americas as part of what Alfred Crosby (1972) called the "Columbian Exchange." From the east came rice and cotton, both raised in the Ottoman Empire, especially in Egypt under a system of intensive irrigated agriculture using slave labor. Initial Islamic objections to the drinking of wine and coffee and to the smoking of tobacco had some effect in slowing agricultural change in the Ottoman dominions, but not as much as might be expected. Imperial viziers came to relish all three of these indulgences. Wine was of course a quintessential traditional Mediterranean beverage. Bottles sealed with cork from the bark of the cork oak (a native of Portugal and Spain) first appeared around 1530 and, along with wooden barrels, began to replace ceramic amphorae and wineskins. The Ottoman rulers relished fine

A typical Mediterranean market in the town of Orvieto, Umbria, Italy, where native Mediterranean fruits and vegetables are sold alongside others introduced after the discovery of the New World. (Photo courtesy of J. Donald Hughes)

wines from Cyprus while it was still in Venetian hands, and throughout the period it was possible for traders to speak of "Ottoman wines" that came from countrysides where they had been treaded and fermented since ancient times. In Venice, the government provided shipbuilding workers in the Arsenal with wine from the seventeenth century onward.

As far as coffee is concerned, its native home was in Ethiopia, it was mainly cultivated in southern Arabia (the scientific name is *Coffea arabica*), and the idea that it was an intoxicating beverage and therefore forbidden by the Qur'an was dissolved in an immense wave of popularity among the Ottoman peoples. Coffee houses were all the rage in Istanbul by the 1560s. Beginning in the 1600s, the same social habit spread across the western Mediterranean, where it also successfully overcame opposition. It was generally brewed as a thick, strong, even muddy-seeming mixture still called "Turkish coffee," although the Greeks insist on calling it "Greek coffee." Its centers of production, however, moved from the Middle East to the tropics of Indonesia (hence the name "Java" for coffee) and to the New World. By 1785, almost all the coffee arriving at the port of

Marseille came from the French West Indies, and much of this was reexported to the Ottoman Empire, reversing the earlier direction of trade.

Tobacco, a crop whose cultivation had been introduced soon after the discovery of the New World through Spain, became enormously popular and was planted in every Mediterranean country, displacing food crops and impoverishing the soil. It was sometimes credited with healing powers and called *yerba santa* ("holy herb"), but others more accurately termed it a noxious weed.

Two members of the nightshade family, the tomato and potato, also entered Mediterranean agriculture in the early modern period. A native of South America that had spread northward to Mexico in pre-Columbian times, the tomato was carried on shipboard to Spain, where it was possibly called *pomo de moro* (Moor's apple), from which the French name, *pomme d'amour* (love apple), derived by similar sound, not meaning. Cultivation of several varieties became widespread in the ensuing decades in Italy, where it was called *pomodoro* (golden apple), a further borrowing of similar sound but corrupted meaning. Today Italian cooking and the ubiquitous *pomodoro* seem inseparable, but the first cookbook to include tomatoes among the ingredients was not published until 1692, in Naples. Its use as food came first in the Mediterranean, since in northern Europe it was regarded as poisonous.

The potato was first mentioned by the Spanish writer of the *Chronicle of Peru*, Pedro Cieza de Leon, in 1553. Potatoes were soon a standard food item on Spanish ships; they kept well and seemed to prevent scurvy. Within twenty years they were planted in Seville, where the archives of a hospital reveal that sacks of potatoes were ordered for provisions. Potatoes grew well in Mediterranean climate and soils, although not in the warmest and driest areas and not as successfully as further north in Europe, and thus they were never to assume as large a portion of the diet as they did in more northerly climes. The wreck of one of the ships of the Spanish Armada in 1588 is supposed to have introduced the potato to Ireland, but other sources claim that the introduction came at the hands of Basque fishermen on their way to the fishing grounds off Newfoundland.

Maize (Mexican corn) was perhaps the most important domestic plant introduced to the Mediterranean basin from the Americas. First developed by seed selection from an ancestral grass in the valley of Mexico, maize was totally dependent on human planting and care for its survival. Before the arrival of Europeans its cultivation had spread north of the Rio Grande, to South America and to the West Indies. Columbus brought seed back to Spain on his first voyage, and from Spain its propagation around the Mediterranean was relatively rapid, especially in Morocco, southern France, northern Italy, and the Balkans. It was common in Tuscany by 1550, and in Lombardy it became the basis of the popular traditional dish polenta. Soon, as noted earlier, it was growing in Egypt next

Maize (corn), a plant first domesticated in Mexico, was adopted in Mediterranean agriculture after the Spanish explorations and conquests in the Americas. It is found growing in the shadows of the pyramids in Giza, Egypt, as it also did in the shadows of the Mexican pyramids. (Photo courtesy of J. Donald Hughes)

to the pyramids. Among the Spaniards who brought it from the New World it was mostly regarded as a food for Indians and animals; Spanish grandees in Mexico would eat it only as an alternative to starvation. Other Mediterranean people did not share this aversion, and by the end of the early modern period maize had become a staple crop. In addition to the multiple uses of the kernels, the stalks and leaves furnished fodder for domestic animals.

Trade was not an occupation without dangers, however. Along with the perils of stormy seas must be ranked piracy of Christian ships in Muslim waters, and the reverse. Treasure convoys from Alexandria to Istanbul were captured, causing the Ottomans to attempt to capture Crete. Barbary pirates sallied forth from North African ports to attack European shipping. But piracy was also conducted between Christian nations; in the seventeenth century, English pirates in the Mediterranean harried Venetian trade as well.

Land tenure varied in different parts of the Mediterranean basin, but there was a tendency for the size of estates to increase, as for example in southern

Italy and Spain. About two-thirds of the population consisted of peasants, and although serfdom gradually disappeared outside the Ottoman Empire, peasants owned little land and were under various forms of disadvantage, including abysmal poverty. They were required to pay rent to their lords and, in some countries, a tithe (10 percent) to the church. Many were bound to a system of sharecropping, common in Italy and southern France, in which the lord allowed the use of land, tools, and draft animals, and in return received a proportion, usually a third or even half, of the harvest. This system discouraged investment by the peasant in the land and equipment, and the effect on the landscape can hardly have been positive.

The tulip craze in the early eighteenth century, when Europeans, especially but not exclusively the Dutch, vied with one another in collecting tulips of the most varied colors and forms and inflated the price of bulbs to astronomical heights, had a brief but extreme effect on those parts of the Mediterranean, especially Turkey, where these perennials grew. Tulip hunters scoured the Anatolian Plateau and Taurus Mountains in search of rare new varieties, turning them into endangered species worth far more than their weight in gold. Tulip traders made fortunes, and then the taste of consumers changed and the market collapsed around 1730.

FORESTRY AND DEFORESTATION

At the beginning of the early modern period, the Mediterranean basin exhibited severe degradation of animal and plant communities, since it had been subjected to millennia of intense human use. Even the forests of the mountain ranges had lost much of their best timber. Agriculture and overgrazing by domestic animals, logging, use of wood for fuel and pitch, and wildfire had entirely altered the ecosystems over the major portion of the region. Forests of anything near the original type survived only in relatively isolated refuges.

The shipbuilding industry of Venice began to exhaust the timber resources of the highlands beyond the marshes that surrounded the city, as Venice struggled both against its old enemy, the Ottoman Empire, and against competition from English and Dutch ships that entered the Mediterranean in significant numbers beginning in the 1570s. For a while, timber came from the exploitation of eastern Adriatic coastlands dominated by Venice, but eventually the former queen of the sea found itself buying ships built in Holland of Baltic timber. Shipbuilding could not use just any tree, but required special woods resistant to water and rot, curved for the construction of hulls, or tall, straight, and strong for masts. A single ship entailed the cutting of between 1,500 and 2,000 oak trees. In the seventeenth century the Dutch outflanked Venetian trade by using northern European wood for shipbuilding, seizing the Spice Islands, and bringing eastern products around the Cape of Good Hope more rapidly and efficiently than the Venetians could manage over the traditional routes through the Near East, where they had to deal with the Ottomans. In the last half of the eighteenth century, the decline of Venice was virtually complete, and the Mediterranean had lost its position as the central carrier of commerce. Similar declines of available forest occurred around other large Mediterranean port cities such as Barcelona, Genoa, Naples, and Messina. The sultan asserted control over cutting of timber in Lebanon to regulate the use of that scarce strategic material, unwittingly following the precedent of the Roman emperor Hadrian, but with many fewer trees to protect.

Forest materials were therefore increasingly valuable. They were needed to build the ships that made exploration, settlement, and exploitation of distant lands possible. Many of the newly discovered continents and islands were rich in forests, so that fine timber could be imported to the colonizing countries, or shipyards could be located abroad. The top-quality forest located along the Atlantic coast of Brazil, for example, provided the Portuguese with shipbuilding timber that was rare or unavailable by this time in Portugal. The sacrifice of those forests may have slowed the exhaustion of the last Mediterranean forests, but at a regrettable cost.

Wood remained a dominant source of energy throughout this period. It was used to keep warm, to cook food, and to smelt metals. Guilds were organized to supply these needs in cities all around the Mediterranean; there were guilds of woodcutters, carpenters, and builders in Ottoman Istanbul (the former Constantinople) in 1638, for example. Prices of firewood in virtually all cities increased throughout the early modern period at rates more than twice those of prices in general. The same was also true in the countryside; as Fernand Braudel famously noted, in Medina del Campo in Spain, wood for the fire was more expensive than the food it cooked in the pot for supper.

In many places, forest removal for the expansion of farmland occurred as the rural population increased and yields of foodstuffs declined on land cleared previously. This was the case, for example, in Sardinia during the seventeenth and eighteenth centuries. Overgrazing and erosion were widespread in the countryside of southern France, Italy, and Sicily during this period. Soil depletion—the loss of chemical nutrients essential to plant growth—is common as a rule in agricultural soils when biomass is simply harvested without provision for adding organic material, unless periods of fallow allow for regrowth of green vegetation that can be plowed under to decompose. Deforestation in the uplands caused erosion that carried some soils to lowlands but often deposited them in the sea, where they were useless unless they accumulated enough to create new land along the seacoast. For example, the major extension of the Tiber Delta into the Tyrrhenian Sea since 1500 has left Ostia, the former port of Rome, a considerable distance inland, and from the ancient ruins there one can no longer see the sea. The land created in this way, however, was a tiny fraction of that lost to erosion in the mountains. It was, in addition, usually marshy and an ideal breeding ground for mosquitoes. The mosquito species that carry malaria in the Mediterranean area prefer open country, so that deforestation undoubtedly had a role in the spread of the disease.

As John R. McNeill points out (1992, 313), a "symbiotic relationship" exists between grazing and erosion, both of which can be damaging to agriculture. Erosion increased the cost of agriculture, leading farmers to abandon cultivation. The only alternative use of the abandoned land was the grazing of goats, which further degraded the vegetation, preventing its regeneration and exacerbating erosion. Some of the resulting landscapes can only be described as badlands.

THE RENAISSANCE AND NATURE

Italy was the setting for an innovative attitude toward nature that appeared at the beginning of the early modern period and helps to distinguish it from the preceding Middle Ages. This view appreciated nature for itself and for its value to human sensibility and use rather than accepting its traditional theological meaning. It was a thoroughly anthropocentric view that rejected, or more accurately, ignored both the contrasting theocentric views attributed in Chapter 3 to Innocent III and Francis of Assisi.

The idea that a human being could and should examine nature with intellectual curiosity for its own sake was not entirely new. Aeneas Silvius Piccolomini (1405–1464), who became Pope Pius II, had an enthusiasm for the natural environment. He loved early-morning walks in gardens and rowing on lakes, and took his cardinals on picnics in groves of trees in preference to dull indoor consistories. He doubted the stories told in bestiaries, such as the one about the barnacle goose born of oysters engendered by the fruit of a tree that had fallen into the water, or that of an herb revealed to Charlemagne as a miraculous cure for the plague.

But humanism soon went beyond simple enjoyment and mild skepticism to bold departures from religious doctrine. Pico della Mirandola (1463–1494) asserted that free will enabled human beings to choose their own nature and set their own limits; placed midway between the divine and the animal, they could choose to be gods or beasts. Doubtful of free will, Leonardo da Vinci (1452–1519) plumbed the other extreme of humanist thought, asserting that necessity is the law of nature, and that nature never breaks her own laws. His life was devoted to understanding the way nature operates, an intellectual endeavor reflected in his drawings of the movements of cats, the patterns of waterfalls, and the position of a fetus in the uterus. Seeking to find natural laws, he investigated every aspect of nature that caught his attention, transcending the divisions between the sciences and the humanities as well. He thought that the insight gained in this way could be used for human purposes as yet unachieved; for example, that an adaptation of the structure of birds' wings could enable human flight. Some of the machines he invented, such as an improved olive press, were intended for practical agricultural and economic purposes. The basic principle that science could inform and stimulate technology was to alter the relationship of humankind and nature, but not until centuries had passed. Some have seen in da Vinci an inhuman intellectuality—how could he so meticulously sketch an autopsied pregnant womb, or calmly devise machine guns and shrapnel bombs generations before they would actually be produced and horribly used? Perhaps he was even more aware of inhumanity than his critics, but willing to look with unflinching eyes on what human nature can bring forth, and present the image because it is true. The utter humanity of his paintings, such as *The Virgin of the Rocks*, reveals his compassion, and it was a compassion that extended to other species as well. "The time will come," he wrote, "when men . . . will look on the murder of animals as they now look on the murder of men."

Niccolò Machiavelli (1469–1527) in his famous works *The Prince* and *The History of Florence* and his lesser-known *Discourses on Livy* discussed the environmental conditions that are favorable or unfavorable to a republic. He noted that a city needs fertile soil, but that an overabundance of it breeds idleness and the love of luxury because cultivation becomes too easy. A harsher environment could spur human effort and creativity through sheer necessity. His reasoning is reminiscent of Hippocrates's *Airs, Waters, Places*, which he had probably read, in which the renowned ancient Greek physician argued that the physical setting and climatic conditions of a city determine the psychological character and health of its inhabitants. In accordance with this doctrine, Machiavelli advised against recruiting soldiers from countries with temperate climates, since such areas do not ordinarily produce men of psychosomatic harmony and courage. Instead, he advised conscripting young men native to the republic, since they would have a love of their own land and therefore a desire to defend it. However, Machiavelli believed that sumptuary laws could instill diligence and energy among the people, counteracting the softening influence of soil that bore crops too effortlessly. A prince has to employ humans against natural forces. And humans have a certain independence from nature, since they can take even deserts and swamps and make them habitable. Thus he ameliorated da Vinci's doctrine of natural necessity with a touch of Pico della Mirandola's assertion of the human ability to exercise choice.

Galileo Galilei (1564–1642) is remembered for his discoveries made through the telescope, but he also spent much intellectual energy studying Earth and the motions of water in the tides and of bodies near Earth's surface. He was condemned by the Church of Rome for embracing the Copernican system, which held that Earth is a planet revolving around the sun, and was criticized for imputing imperfections to the heavenly bodies, since he reported mountains on the moon and spots on the sun. But even more threatening to scriptural religion was his insistence that universal laws could be discovered by observing the movements of the stars and planets, falling bodies, and the level of the sea. He was wrong about the mechanism that causes the tides—he thought it had to do with the inertia of Earth's rotation rather than the gravitational attraction of the moon and sun—but the principle of induction from observation was one of the prime roots of the Scientific Revolution, a new paradigm of the early modern period that would radically alter human perceptions of the natural world.

Humanism and science thus received a tremendous impetus from Mediterranean minds at the outset of the early modern period. This impetus did not continue there, but moved northward, returning to the European Mediterranean lands only after its further development in northwestern Europe. The southern and eastern margins of the Mediterranean basin, where classical texts and philosophical speculation had survived through the earlier Middle Ages, remained virtually untouched by this new thought.

CONSERVATION

The old royal custom of reserving forest lands for hunting by kings and nobles persisted into the early modern period. Ferdinand of Aragon, for example, enjoyed the chase in his deer parks so much that he became a collector of "books of sport" such as the manuscript *Book of the Hunt* by Gaston Phèbus, prince of a Pyrenees county in the fourteenth century. The hunt was a way of life for many aristocratic men of the age, and they desired to protect not only the wildlife that made their sport possible but also the patches of habitat that served as indispensable environment for beast and hunt alike. Royal power and wealth remained great enough to make this possible. But there were other, perhaps more pressing, reasons for forest conservation.

One of the measures suggested by the many demands on forests, especially those for navies, was the reservation of forest lands for purposes such as shipbuilding. Venice tried this method in northern Italian forests, but not successfully due to resistance from villagers, private shipbuilders, and peasant laborers. Venice was unable to protect the forest reserves because it lacked sufficient police manpower to protect them, and lost interest as they were degraded.

The French Forest Ordinance of 1669 applied conservation measures throughout France, which of course included the Mediterranean sections. It turned the forests into a branch of the state-managed economy and attempted to ensure a dependable supply of timber for the navy. In so doing, it put France in the forefront of the development of the art and science of forestry and set the stage for the great forestry schools of the nineteenth and twentieth centuries. It restricted the use of forests for charcoal and in various woodworking occupations; prohibited the grazing of cattle, sheep, and goats in forests; and reserved certain important trees as seed sources. To some extent the measure was intended to protect the forest resource from destruction as a result of the incipient Industrial Revolution, and to conserve the watersheds of rivers in drier sections such as the Pyrenees. For the greater part, the ordinance was enforced, and the survival of French forests, including those in the Mediterranean districts, was better than elsewhere.

Conservation of another renewable natural resource is represented by the regulation of fishing along the French Mediterranean seacoast and the waters accessible to it. This was in the hands of a group of *prud'homies* (guilds) of fishermen that regulated fishing schedules and grounds, settled disputes, prevented competition, and had the effect of preserving a variety of fishing methods (Pavé 2003). These guilds first appeared in Marseille in the Middle Ages, and they expanded rapidly in the eighteenth and early nineteenth centuries. Their power was confirmed by the Royal Council at various times throughout the eighteenth century and was not interfered with by public authorities even during the French Revolution and its aftermath. This self-regulating industry prevented overuse of the resource as late as 1850, and was paralleled by similar institutions in the Spanish Mediterranean and possibly elsewhere.

TECHNOLOGY

The inventions that changed the world in the early modern period, including the compass, gunpowder, and printing, were made in China, as Francis Bacon perceptively noted in 1620. It is evident that the first two, along with the astrolabe (an astronomical instrument used to determine latitude), helped to make possible the European movement to control large areas of the world. Gunpowder also enabled the Turks to conquer Constantinople in 1453 and thus to gain a

Astrolabe used by a sixteenth-century astronomer, dated 1569. An astronomical instrument used to determine latitude, the astrolabe helped to make possible the European movement to explore and to control large areas of the world. From the Paris Musée des Arts et Métiers. (Bettmann/Corbis)

capital for the Ottoman Empire, which dominated half of the Mediterranean basin throughout this period. Gunpowder required potassium nitrate, sulfur, and charcoal. Potassium nitrate and sulfur were obtained from mining, with attendant damage to the landscape, and charcoal was derived from wood, adding another important use to that forest product and increasing the pressure toward deforestation.

Printing is, however, probably the most important invention to receive widespread adoption during this period. Block printing is known from China in the ninth century, and movable type from the eleventh century. Although there was a Mediterranean precedent of movable type used on clay (the Phaistos Disk) in Crete during the second millennium BC, it was apparently unique and had no subsequent influence. Johannes Gutenberg printed books using movable type in the 1450s in Mainz, Germany. The first printed book in Italy, from Subiaco near Rome, bears the date 1464, and books were printed in Venice five years later using for the first time the standard typeface with upper and lower cases. Aldus Manutius, a Venetian printer, introduced Italic letters around 1500. Printing entered Spain at Valencia in 1474. The first Turkish printing press did not appear until 1729, near the end of this period, since before that time the Ottoman authorities had forbidden the printing of any books in Turkish or Arabic (although Jewish, Armenian, or Greek presses could be permitted in Ottoman cities). The press just mentioned was destroyed in 1742, and printing was not resumed until 1784. European presses, however, had been printing Arabic books since the sixteenth century. Printing had as its most powerful effect the exchange of ideas, with worldwide impact on the relationship of human economy to the natural environment. It also had direct influence on the use of resources such as lead and paper. During the early period of printing, paper was made from rags, and at the time Italy and Spain led in paper manufacture and exporting, but they lost this lead as paper factories became established in the northern European countries. The transition to paper manufactured from wood pulp, with grave import for forests, did not occur until after the end of this period.

Another crucial invention that came into its evident power during this period was the mechanical clock. There had been clocks during the Middle Ages that tolled the hours from steeples and clock towers, but in the early modern period they gradually became ubiquitous, and could tell minutes as well as hours. This development changed attitudes toward work and production. Also of great importance was the fact that an accurate clock on shipboard was essential to the determination of latitude. An astrolabe could tell a navigator how far north or south the ship was by observing the elevation of the sun above the horizon, but in order to find the position east or west, it was necessary to discover the difference in time between the ship and a known point. Through most of this period, clocks were not dependable enough to accomplish this because they gained or lost time due to friction, temperature changes, and the motion of the ships. During the eighteenth century, Thomas Harrison experimented with a series of new models that improved balance, reduced friction, and avoided the effects of temperature by using a combination of metals. By the time Captain James Cook was exploring the Pacific in the 1770s, it was possible for him to establish longitude with some accuracy. Similar clocks were soon in use in the Mediterranean, and ended for good any necessity for large ships to hug coastlines in navigation, since they could confidently sail across the sea directly to their destinations. The barometer, thermometer, and microscope were also invented and set the scene for the scientific revolution with its vast opening of human understanding of the natural world. Still another invention, the telescope, dramatically expanded human perceptions of the cosmic environment, although the condemnation of Galileo by the papacy in 1633 is a reminder that it was not just the Ottoman Empire that resisted modernization.

INDUSTRIES

Metallurgy in general, and especially the smelting of iron in the blast furnace, was perhaps the most damaging force upon the landscape, since not only did mines and spills of excavated material and slag produce scars, but also the consumption of the forests for fuel was enormous. Although the use of coal began in this period, coal could not yet economically be substituted for charcoal in the smelting process, and in any case the Mediterranean region is poor in coal deposits. Although the use of iron had not yet reached its peak, iron was already the predominant material for tools, machinery, and armaments. In the Mediterranean area, ironmaking centered in the Basque region of Spain, Catalonia and the Pyrenees, Tuscany, Genoa, and Corsica, with some centers also in southern France, and there were few areas without at least some small ironworks.

The result of wood utilization by metallurgy in iron and other metals such as lead was the loss of wild forests, which were consumed as fuel and used as material for mining supports and machinery. In mountainous southern Spain, a district of heavy mining and smelting, there was major use of trees. Since there was not much forest there at the beginning of the period, and the population was relatively crowded, very little forest remained in the eighteenth century.

Pollution was a widespread result of industrial activity. Water pollution came from mine drainage and the wastes from ore processing, and because these often occurred in mountainous areas, the tainted water flowed down into inhabited and cultivated districts before reaching the almost landlocked sea. Smelting operations polluted the air and caused complaints even where high chimneys had been constructed to dissipate the smoke.

In the sixteenth century, under the older industrial system, the Mediterranean saw rapid economic growth, but it was not to continue. The "putting-out" system, in which the merchant advances to the artisan the raw materials and money for wages and then sells the finished product, had spread widely in the Mediterranean and undercut the medieval guilds. The Industrial Revolution was slow to come to the Mediterranean basin. Indeed, by the second half of the eighteenth century, while factories and mechanization were proliferating in England and northwestern Europe, Mediterranean industries such as textile production in northern Italy were declining due to competition with cheaper goods from the northern lands, wars, and depletion of resources. The Italian wool industry collapsed, as did the weaving of silk, although production of raw silk continued there and in Crete and southern France, where it had become established earlier. The once-vigorous Spanish cloth industry had already failed due to guild restrictions that did not allow adjustment to new commercial circumstances and to oppressive royal attempts at control. The Mediterranean basin as a whole entered the nineteenth century at something of a loss as far as industry was concerned.

DISEASE AND EPIDEMICS

The Black Death did not disappear after its ravages during the Middle Ages. It became endemic, remaining present in rodent populations and recurring in outbreaks from time to time, as in Verona in 1510. Its persistence depended on environmental conditions.

Plague usually attacked in periods of high population when overcrowding led to unsanitary conditions and therefore to close human contact with the rodents and the plague-transmitting fleas they carried. Although unaware of the mechanism of infection, Italian cities enacted public health measures including preventive quarantine and street cleaning. These attempts to counter the disease by cleaning up the urban environment—removing piles of refuse—were sometimes frustrated by riots of destitute ragpickers who objected that they were being deprived of a source of their subsistence. During the seventeenth century, plague had an endemic focus in Syria, Turkey, and the Balkans, and serious episodes occurred elsewhere in the Mediterranean. In 1656, Italy was ravaged by an epidemic of plague, and physicians who dared to treat its victims approached them wearing thick robes and masks with long birdlike beaks that they stuffed with flowers in an attempt to filter out the contagion. These measures may have helped to avoid the pneumonic form of the plague that could be transmitted on moisture droplets through the air, but would have had little effect in fending off fleas. The disease reached Malta for the first time in 1675 and carried off 8,732 victims, 15 percent of the population. There were several epidemics in the eighteenth century, including one that reached Marseille from Syria in 1720 and wreaked havoc on Provence for two years, and another in Venice in 1743. The Napoleonic campaigns in Egypt between 1798 and 1801 resulted in outbreaks of plague among French and Turkish soldiers and the British navy.

As the western terminus of the Silk Road from North China and the sea route from India through the Persian Gulf and Red Sea, the eastern Mediterranean had long received epidemics from the east, but in the early modern period the import of infections from the lands across the Atlantic also began. Of course it is only fair to point out that Mediterranean Europe exported more diseases by far than it received from the New World.

Syphilis seemed a new and strange disease to the Mediterranean peoples when it infected so many of them in the 1490s. They wondered where it came from: the Spaniards called it *morbus gallicus,* the "French disease," and the French called it *mala napolitana,* the "Neapolitan disease," since the army of Charles VIII had contracted it during a campaign against Naples in 1494. As an Italian author named Moscardo remarked, "Not knowing whom to blame, the Spaniards call it the French disease, the French the Neapolitan disease, and the Germans the Spanish disease" (Gould 2003, 193). Francisco López de Villalobos (1473–1556), a professor of medicine at the University of Salamanca, was the first doctor in Spain to write about the malady, in 1493–1495. He speculated whether it might have been contracted by the sailors of Columbus from Carib women in the West Indies and brought back to Spain. Girolamo Fracastoro (1478–1553), a medical graduate of the University of Padua who also studied botany and geology, among other subjects, first called it syphilis. He took the name from a mythical shepherd, Syphilus, who was stricken by Apollo's arrow for worshiping his king, Alcithous of Ophyre (Haiti), instead of the god. This arcane legend was possibly derived from the story of Sipylus, one of the children of Niobe who were slain by Apollo because their mother had boasted that she had more children than Leto, Apollo's mother (Marks and Beatty 1976, 120–121). Fracastoro noted the method of transmission of the disease through sexual intercourse and other intimate contact. He gave support to the idea that the disease came from the New World by reporting that a medicine effective against syphilis was derived from guaiacum, a sacred tree of the West Indies associated with a grove of a deity whom he identified as Urania, goddess of the air. Actually guaiacum proved ineffective against syphilis, but mercury, although poisonous, did have some curative power.

Typhus was also first noted in the 1490s, but came from the east rather than the west, since it first came to Spain from Cyprus. Carried by lice, it is a disease of environmental conditions such as crowding and poverty. Typhus epidemics occurred every few years in Italy, and doubtless in other countries around the Mediterranean throughout the early modern period. Many of the outbreaks were associated with war.

Malaria was another recurring disease of the period. Its name means "bad air," but its real vector, mosquitoes, was still unknown. In 1602, some 40,000 deaths were recorded from it in Italy, and in the eighteenth century malaria came to Rome every summer, so those who could afford to do so abandoned the city and fled to towns in the mountains during that season, a custom that dates back to antiquity. Pandemics swept across the Mediterranean lands in the sixteenth and seventeenth centuries, and at least four times during the eighteenth century. The use of an infusion of the bark of a Peruvian tree called cinchona, which contains quinine, to suppress the symptoms of malaria became known in 1632 and was widely used by the 1650s, but a dependable supply of this medical substance was not available for two hundred years afterwards.

A pandemic of smallpox hit Mediterranean populations in 1614. Interestingly, inoculation against smallpox existed as a folk practice in various parts of the Ottoman Empire. It was a custom of Greek peasant women in Thessaly and the Peloponnesus, for instance. Word of this prophylaxis, and the practice itself, spread to other parts of Europe and the Mediterranean beginning in the 1720s. By the end of the century, the effectiveness of vaccination with cowpox had received scientific demonstration.

Only the diseases that receive the most comment by contemporary writers are mentioned here. There were many others that resulted in part because of environmental conditions of exposure. There were six instances of diphtheria epidemics in Spain in the late sixteenth and early seventeeth centuries, to give one example. Many other events of this type surely remain unrecorded, especially on the southern and eastern margins of the Mediterranean.

CONCLUSION

The early modern period in the Mediterranean basin gives a foretaste of some of the sweeping changes that were to affect the human relationship to the natural environment in the following two centuries, although those living at the time might have had difficulty seeing just what they would turn out to be. The spread of iron production, accompanied by a major increase in the as-yet-small output of steel, foreshadowed the advent of the Industrial Revolution, which was already stirring in England and other northern lands in the second half of the eighteenth century. Science, which had its first stirrings in Italy as much as any other land, had begun to shake not only ways of thought, provoking great resistance in the establishment, but also ways of doing things, which would combine with new technological inventions to transform agriculture, transportation, and industry. Resources were being consumed at a greater rate, ores mined, forests felled, and land depleted more rapidly than in preceding periods, but all that was but an inkling of the massive changes to come.

One historical development of great importance during the period was the opening of direct sea trade to the east by Vasco da Gama and other Portuguese seafarers, and the discovery of hitherto unknown lands to the west by both Spanish and Portuguese explorers. The greatest changes as a result of the western venture were felt by the human inhabitants and ecosystems of the Americas and the Atlantic islands, but the "Columbian Exchange" went both ways, and the Mediterranean would never be the same again. The most important immigrants consisted of a host of new domestic plants, including not only maize, potatoes, tomatoes, and tobacco but also sweet potatoes, several kinds of beans, squashes, pumpkins, chilis, and a number of unwelcome weeds. There were also the products of other plants too tropical for the Mediterranean environment, notably chocolate and vanilla. Few if any humans from the New World came to settle in the Mediterranean, since deaths from disease were overwhelmingly frequent and frighteningly rapid among those who came on returning ships, whether by choice or by force. Very few other animals made a successful transatlantic crossing in the permanent sense from west to east; the peoples of the Americas had few domestic animals. Among those that arrived in the Mediterranean, however, were the turkey and the guinea pig, obviously renamed in their new homes, the turkey after a Mediterranean country and the guinea pig after a West African coast, although both came from America. Among diseases, only the syphilis spirochete spread and was feared for good reason in the Old World. The exchange was not concluded, however, for good or for ill.

This period saw growth in the size and relative importance of cities beyond the Middle Ages, an increase in commerce and industrial production, and a flowering of science, but in each of these factors the Mediterranean began the period as a leading and central area relative to its neighbors and ended it lagging behind northwestern Europe. In any case, it must be pointed out that agriculture was still the chief economic activity affecting the environment and the one involving the vast majority of the inhabitants of the Mediterranean world.

References

Braudel, Fernand. 1972. *The Mediterranean and the Mediterranean World in the Age of Philip II*. New York: Harper and Row.

Crosby, Alfred W., Jr. 1972. *The Columbian Exchange: Biological and Cultural Consequences of 1492*. Westport, CT: Greenwood Press.

Gould, Stephen Jay. 2003. "Syphilis and the Shepherd of Atlantis." In *I Have Landed: The End of a Beginning in Natural History*, 192–212. New York: Three Rivers Press.

Marks, Geoffrey, and William K. Beatty. 1976. *Epidemics*. New York: Charles Scribner's Sons.

McNeill, John R. 1992. *The Mountains of the Mediterranean World: An Environmental History*. Cambridge: Cambridge University Press.

Pavé, Marc. 2003. "History of the Sustainable Management of the South-West European Fishing." In *Dealing with Diversity: Proceedings of the 2nd International Conference of the European Society for Environmental History*, eds. Leoš Jelecek et al., 87–90. Prague: Charles University, Faculty of Science.

5

THE MODERN ERA
(1800–1959)

The green mantle of the Mediterranean is . . . composed of many facets. There is the original forest, grassland, and desert. There are the imports, some dating from classical times or even before, others made within the last few decades. There are representatives of both these types cultivated for food, clothing, or other commercial purposes; and finally there are the ornamental plants of all kinds that have been nurtured and developed solely to delight the eye of man. The presence of all these forms of plants in one comparatively small region gives the Mediterranean an historical, intellectual, and aesthetic interest . . . which can be matched nowhere in the world.

—RICHARD CARRINGTON, *THE MEDITERRANEAN*, 1971

Man is everywhere a disturbing agent. Wherever he plants his foot, the harmonies of nature are turned to discords.

—GEORGE PERKINS MARSH, *MAN AND NATURE*, 1864

The modern era saw the rise of the Industrial Revolution, with its unprecedented impacts upon the natural environment. This radical change in the technology and organization of the manufacturing industry began in northwestern Europe in the late eighteenth century and spread to parts of the Mediterranean economy in the nineteenth century. A new order of production was originated that was no longer predominantly agricultural, although the majority of the working population may still have been employed in agriculture for most of the period. The widespread advent of the use of machinery made of iron and other metals, and the harnessing of power from nonrenewable fossil fuels, first coal and later natural gas and petroleum, made possible a transformation of the landscape of a degree and kind never seen before.

It may also have aided the northern European nations, with better economies and weapons, to exert their power in the Mediterranean at the expense of the less industrialized south until the later part of the period, and to take commercial advantage of Mediterranean resources and raw materials. The British had seized Gibraltar in 1704, and retain it to the present day. At various times, for varying periods, the British took control of Malta (1814), the Ionian Islands, Cyprus, Egypt, Palestine, Transjordan (later called Jordan), and Iraq. The French, having already annexed Corsica, extended their aegis over Algeria (1830), Tunisia, Morocco, and Syria during periods of various lengths. They tried to keep a finger in the Egyptian pie, briefly occupied that country militarily under Napoleon, and constructed the Suez Canal in 1869. Spain also established a protectorate in northern Morocco. The Austrians and Hungarians annexed Trieste, Slovenia, and Croatia, including the Dalmatian coast, until World War I fragmented their empire. The Italians, after their industrialization had begun and their unification achieved, engaged in an imperial operation in Tripolitania (Libya), and occupied the Dodecanese Islands between the world wars. The period saw the weakening and shrinkage of the Ottoman Empire, which first lost its North African dominions to local rulers and then to European powers, and at the end of World War I saw its Arabic possessions in the Near East taken by France and Britain, while the Turkish homeland became a republic under Mustafa Kemal Atatürk (1881–1938), who defeated the postwar invasions of the Greeks and other Europeans and carried out a series of modernizing reforms, including some seeking environmental restoration. Realizing the need for restoring the deforested and eroded sections of Turkey, he expressed a desire for a religion with tree planting as its chief rite.

Among the sweeping changes brought about by the Industrial Revolution were many that affected the natural environment. In its initial phase, when wood and charcoal were still the major sources of energy, deforestation threatened the remaining tracts of forest. With the shift to fossil fuels, deforestation may have slowed, but a vast increase in mining and manufacturing activity generated pollution that sullied the elements of earth, air, and water, and to that list must be added the urban areas. The population in the Mediterranean area and in much of the rest of the world began an accelerating upward spiral, using land and water, swelling cities, and producing unprecedented flows of garbage and sewage. The pollution of the Mediterranean Sea itself, and the possible depletion of its fish and other aquatic resources, became subjects of concern. Many forms of life were threatened with extinction, and the quality of human life changed in countless ways for better and for worse.

HUMAN SETTLEMENTS: URBAN GROWTH, POPULATION INCREASE

Two major changes appeared in human settlements in this period: a huge increase in population and an increase in size of cities. The escalation of population was due to declining death rates; famine crises decreased in frequency but did not entirely disappear in the nineteenth century, and improvements in public sanitation and medicine lowered the mortality from diseases. Birthrates followed different trends in different parts of the region, rising in some Mediterranean countries (mainly in the south and east) and declining elsewhere after the middle of the nineteenth century, especially in the most prosperous countries, such as France. The total population of the Mediterranean countries more than doubled during the nineteenth century, but the urban population grew more rapidly than that. There were more large cities, and those that already were large increased proportionately in size. Among those that expanded significantly were the French port city of Marseille and the Iberian capitals of Lisbon and Madrid. In Italy, the cities that participated in the Industrial Revolution enlarged rapidly: the port city of Naples and the northern industrial center of Milan each had about a million inhabitants in 1881 and two million in 1931. Italy experienced an overall increase of 2.4 times between 1800 and 1931, from 17 million to 41 million, in spite of a massive emigration to the United States and other countries during that period. Meanwhile, the cities of the agrarian south increased more slowly or, like Potenza and Syracuse, actually shrunk, as workers moved to larger cities where there were more jobs. All around the Mediterranean, the pattern of rural immigrants moving to urban centers continued, but by the early twentieth century the population of cities was itself growing due to improvements in public health.

The earlier population of the Ottoman Empire must be estimated because no real censuses were taken until the 1880s, and comparative figures are misleading, since the empire shrank in area during the entire period. In 1914 there were twenty-six million inhabitants. Important Ottoman cities were located on caravan routes or at major harbors where products were transferred from land transport to maritime trade, or the reverse. The plans of these cities were irregular, following patterns that trace back to earlier eras. Istanbul, the imperial capital, was by far the largest, with 750,000 in population in 1800, increasing only to slightly over a million in 1924. During that period its population was highly diverse, with large Greek, Armenian, and Jewish minorities living in more or less well-defined quarters that were to some extent separated by walls and gates. Houses were built around central courtyards and were therefore inward-looking,

The entrance to the Bosporus Strait in Turkey, viewed from Istanbul's Topkapi Palace across the Golden Horn. A modern bridge spans the water from Europe, on the left, to Asia, on the right. (Photo courtesy of J. Donald Hughes)

a pattern that had been characteristic of the entire Mediterranean in previous centuries. The attempts of the Ottoman government to initiate Westernizing reforms in the nineteenth century found expression in more regular street plans; these were enabled to some extent by the all-too-common fires that destroyed older sections of the cities. Health and sanitation improved; there were outbreaks of bubonic plague in the early nineteenth century, but hygiene measures were enacted and hospitals were built later in the century. Over a million Ottoman subjects, most of them Christians, emigrated to America from 1860 to 1914, but Ottoman cities still grew, at least those that were ports: Salonica and Izmir almost tripled in size between 1800 and 1914, while Beirut mushroomed from 10,000 to 150,000.

After World War I, cities in the Mediterranean grew even faster, spreading outward over former agricultural land in an unprecedented surge of suburbanization. Between 1917 and 1937, Cairo grew from 800,000 to 1.3 million, spreading to the islands in the Nile and across the river in the direction of Giza.

Infrastructure improved, although it was hard-pressed to keep up with the growth. Water and sewage systems, gas, electricity and telephone services, buses, tramways, and motor traffic increased during the 1920s and 1930s.

AGRICULTURE AND PASTORALISM

Very important in the modern era was a new agricultural revolution involving the mechanization of agriculture that appeared first in northern Europe. Planters, harvesters, and tractors would not only save hand labor and result in a sizeable increase of food production but also change the pattern of fields to accommodate the machines, a force for consolidation of ownership over larger tracts of land—and one of the reasons for the abolition of serfdom in country after country. Mechanization of agriculture was difficult in the Mediterranean due to soils, topography, and economic depression, and also to a pattern of absentee landownership in which the owners lacked interest in the technological and economic improvement of their lands and tenants. The new agriculture therefore spread very slowly, in many places not arriving to any major extent before the twentieth century. Even by the 1960s, Portugal, Spain, and Greece still lagged far behind the rest of Europe in mechanization, and Italy, although closer to northern nations, still ranked below all of them (France, with half its country in the north, cannot be counted as a Mediterranean nation in this comparison). When it did come, mechanization enabled expansion of the cultivated land at the expense of forest land and open country. Some of this occurred in the years before World War II; Mussolini, for example, succeeded in draining the malaria-infested Pontine Marshes south of Rome and opening the rich soils of the area to agriculture and new towns, a project attempted with negligible results several times before in history from the time of the ancient Romans.

Compared with northern Europe, a larger proportion of the workforce in Mediterranean countries remained engaged in agriculture during the modern period. In the 1920s and 1930s, more than 50 percent of the laborers in Italy and Spain were active in the agricultural sector, whereas in the Netherlands, Germany, and Belgium the figure was under 25 percent. Moreover, productivity was lower in the south. Mediterranean diets in the mid-1950s were high in carbohydrates and low in animal protein; the daily calorie total per capita from meat, fish, milk, and eggs in Mediterranean Europe was 250; in northern Europe it was 940. Meanwhile, the families of Portugal, Spain, Italy, and Greece were spending 40 percent to 50 percent of their incomes on food, contrasted with 30 percent to 35 percent in northern Europe. Undoubtedly the picture was worse in the southern and eastern parts of the Mediterranean basin.

Olive oil press in Dolcedo, Italy, ca. 2000. Olive presses of a more primitive type have been used since ancient times. Leonardo da Vinci invented an improved olive press. The type shown here is seen in many countries around the Mediterranean today. (Owen Franken/Corbis)

One technological improvement that affected a traditional Mediterranean product was the modern hydraulic press used to extract oil from olives. This enabled the production of a much larger volume of high-quality edible oil, since the oil could be expressed from the marc, or pulp, much more rapidly and thus be preserved from rancidity and contamination. In the early twentieth century, Spain was the most prolific producer of olive oil, being responsible for almost 40 percent of the Mediterranean total in 1937. But Italy, particularly the region of Tuscany, along with Provence in France, made some of the most highly esteemed olive oils. Italy produced the second-largest amount, at 24 percent of the total, and Greece was third at just under 15 percent. In order to merit the title "virgin oil," the fruit is supposed to be handpicked and pressed gently, but much of the oil that bears that designation in modern times is in fact a high quality of machine-produced oil. "Extra-virgin" oil commands even higher prices.

In the Ottoman Empire, where industrialization made only minor inroads, at least 80 percent of the population remained agrarian, engaged in production of crops such as tobacco, rice, cotton, grapes, cereals, olive oil, sugar, oranges, and dates. In some parts this was dryland farming using plowing by oxen, but in a few districts irrigation agriculture was conducted as well. Although the Land Law of 1858 defined most cultivated agricultural land as belonging to the state, private owners could obtain a title enabling them to full use of it and the ability to transmit it to their heirs. The result in the empire and in the independent or western-dominated nations that became its successors was widespread owner-ship by absentee landowners who used landless laborers or sharecroppers to farm the land. Labor was intensive, technology was inefficient, and taxation was crushing—taxes on agriculture represented 40 percent of all taxes in the empire, and this does not include customs charges on agricultural exports. Egypt was a special environmental region; it is discussed later in this chapter and in the chapter on case studies. Turkish and Islamic refugees from former Ottoman territories in the Balkans were resettled on the Anatolian Plateau, in the Jordan valley, and in upper Mesopotamia, expanding the agricultural zones there. Iron plows and steam-powered threshers appeared in Anatolia in the 1880s, and irrigation was extended in North Africa and Iraq, but agricultural technology still lagged behind the European portions of the Mediterranean basin.

Greece presents an interesting case in which the state attempted to develop agriculture based on small farms, with only mixed success. At the beginning of the period, industrial agriculture had not yet arrived and population had not in-creased to a great extent. The Greek War of Independence in the 1820s dispos-sessed Ottoman landlords of large estates in the Peloponnesus, which were taken by the new Greek government and made a category of public land called "national estates." Attempts to transfer this land to private farmers succeeded only gradually because wars had decreased peasant numbers, and intensive la-bor that was still required because mechanization had not yet appeared was not initially available. Even so, about 80 percent of the rural population consisted of members of families that owned at least some land by 1881. From that year through World War I, Greece gained considerable territory in the north at the expense of the Ottoman Empire, and in the islands. Large estates emerged in these new lands, but the situation changed with the Greco-Turkish War of 1921–1922 and the "exchange of populations" that followed, bringing 1.625 mil-lion ethnic Greek refugees to Greece. This was a crisis that led the Greek gov-ernment to implement an extensive land reform that distributed about 2 mil-lion hectares (1,265 square miles) to 319,000 families. The average family plot awarded was therefore not large—about 6.25 hectares (15.5 acres)—and was usu-ally split into several disconnected parcels. The Greek countryside was rela-tively depopulated by World War II, the Communist insurrection that followed, the movement of population to Athens and other cities, and the emigration of workers to western Europe. Foreign aid and efforts by the Greek state replaced the agricultural infrastructure that the fighting had destroyed, however, and by 1959 the nation was self-sufficient in cereal production.

Introduced plant diseases constituted a threat to agriculture. The vineyards of the Mediterranean received a devastating blow after 1863, when an aphid known as phylloxera appeared in France and subsequently spread to virtually every wine-producing district in France and around the Mediterranean. It was later realized that the organism had been carried on infected plants from the eastern United States. This tiny but very prolific insect attacked the roots of grapevines, making the grapes useless for wine and eventually killing the vines. It soon became clear that unless a cure could be found, the wine industry of the Mediterranean would virtually come to an end. The effects were indeed disastrous, and some wine-growing districts such as the Alpujarra in southern Spain were depopulated as a result. Vintners tried flooding the vineyards and fumigating them with poisonous carbon disulfide, with only partial success. Then, since it had been observed that American vines were resistant or immune to phylloxera, plant scientists successfully grafted the European grape varieties onto American rootstock, and in 1881 a conference of French winegrowers approved the replacement of susceptible vines with the new grafted ones, rescuing the production of the staple beverage and its prized vintages. Similar methods saved vineyards all around the Mediterranean, including Spain, the Maghreb, the Balkans, and Greece. Because phylloxera preferentially attacks roses, it has become a custom to plant roses in or near vineyards as a kind of early warning system or "miner's canary" to indicate when the aphids are becoming active, so that measures can be taken to protect the vines.

FORESTRY, DEFORESTATION, AND EROSION

During the modern era, there was a wide variation in the condition of forests in the Mediterranean countries, and of the ways in which they were managed or not managed. France established the National School of Waters and Forests at Nancy in 1824, which produced trained foresters who tried to establish scientific forestry regimes in the French Mediterranean districts and in the major northern part of the Maghreb that was controlled by France for much of this period. They had mixed success in suppressing incendiary fires and excluding grazing herds. British efforts in Cyprus were similarly frustrating for the forest managers.

Thus the general picture was not positive. The peoples of many parts of the Mediterranean basin suffered scarcity and deprivation due to deforestation, not to mention flooding, erosion, and siltation. Writing in 1864 in Italy, George Perkins Marsh recorded that "upon the borders of the Mediterranean, the destruction of the forest and all the evils which attend it have gone on at a seriously alarming rate" (191).

In 1868, the forest cover of Sardinia was only 12 percent, of southern Italy 9 percent, and of Portugal 4 percent. Clearing was especially notable in Provence and southeastern France generally, where woodcutters took the trees, and goats and sheep finished the consumption of the vegetation. In the early nineteenth century, eyewitnesses recorded that peasants there were reduced to baking their bread by burning cattle dung and to keeping warm in winter by moving in with the cattle. Spain, southern France, and Italy were importing timber from the Baltic Sea to supplement the inadequate Mediterranean supplies. In the Pindus Mountains of Greece, deforestation led to erosive soil loss that made crop growing impossible in some districts, so that villages were abandoned, but there was little or no regrowth of trees on the deserted land. Other regions with extensive deforestation, but where estimates of cover are lacking, include Turkey and the Maghreb.

The rate of deforestation increased significantly with the harnessing of steam power for sawmills, invented in the mid-nineteenth century, but not used widely in the Mediterranean basin until the end of that century. Steam railroads required millions of trees for railroad ties, while they improved the transportation of logs from forest to sawmill; the Mediterranean had never been well supplied with rivers large enough to float logs. Indeed, because of intermittent flows, cut logs piled up in river bottoms, waiting for the rains and the return of enough water to float them downstream. Studies in central Italy based on land surveys for tax purposes made in the early nineteenth century compared with aerial photographs of the 1950s indicate major changes in the landscape involving about half the entire territory, including a significant enlargement of arable land, and deforestation somewhat balanced by plantations of single species of trees and the spontaneous regrowth of forest on abandoned land. In the southern province of Basilicata, more than 30 percent of the forest was reduced to cultivation or grazing land between 1826 and 1950. By 1900, only about 100 square kilometers of natural forest survived in Palestine, segmented into small isolated groves, and these were further decimated as both sides, the Turks and the Allies, needed ties for the railroads they constructed during World War I and the years afterward.

Plantations of single species of trees, a practice called monoculture, became widespread in the European Mediterranean in the early twentieth century, and in Israel after independence. The practice is intended to produce timber for commercial use; only one kind of tree is planted in a given area to make harvesting easy. The species chosen are those adapted to the climate and relatively fast-growing to reduce the length of the maturing cycle. Among those used are some native Mediterranean pines, the Canary Island pine, and the very rapidly growing Monterey pine (*Pinus radiata*), native to a small section of California that has a Mediterranean climate. Conservationist groups in Israel called the new plantations there "pine deserts" because few native species survived within them. Arizona juniper has been used in northern Greece and elsewhere to create forest cover quickly. Very popular for a time were several *Eucalyptus* species from Australia, now seen so widely around the Mediterranean that they almost seem to be a characteristic feature of the landscape. Unfortunately eucalypts are also extremely combustible, exude chemicals that kill other vegetation in their vicinity, and mostly produce inferior wood in the Mediterranean region; consequently some countries have passed laws forbidding their use in plantations. Plantations are often counted in the surveys of forest cover, but they do not represent a

A huge deposit of boulders brought down by the Laphystas torrent in Pieria, Greece, the result of erosion in the deforested headwaters of the stream. (Photograph courtesy of the Goulandris Natural History Museum, Kifisia, Greece)

restoration of the original forest in any sense of the word and should therefore be listed separately.

It is now generally recognized that erosion is made far worse by deforestation, and in many cases erosion occurs where there would not have been any appreciable amount of it if the trees had not been removed. This idea was known to ancient Greek and Roman writers, but its modern acceptance is to some extent the result of the work of George Perkins Marsh, an American with considerable experience in land use as a sheep owner, lumberman, quarryman, and developer. He was three times elected a member of the U.S. Congress, then appointed U.S. minister to Turkey. He traveled widely in the Mediterranean and finally was made ambassador to Italy by Abraham Lincoln, an office he held from 1861 until his death in 1882, the longest term ever served in one post by an American ambassador. In his influential book, *Man and Nature*, Marsh described many examples of deforestation leading to damage: in mountain soils washed away by heavy Mediterranean rainstorms when there were no trees to break the force of rains striking the ground, and no tree roots to hold the soil together and act as a sponge; and also to farms and towns in the lowlands as floods swept down on them bearing rocks, mud, and boulders. He made careful attempts to quantify the extent of the damage. Not everyone believed him; when he talked with Italian farmers, he discovered that they blamed the storms, not the bared mountainsides, for the damage. The social cause of deforestation and erosion, Marsh believed, lay in the devastating subjugation of the peasants by landowners and oppressive governments, which compelled them to abandon agricultural land that they had won by their labor from the forest. Thus the soil had neither the protection of forest cover nor the care of farmers who knew how to treat it well (Lowenthal 2000, 283).

In the years just before World War II—1938 and 1939—Walter Clay Lowdermilk, assistant chief of the new U.S. agency soon to be called the Soil Conservation Service, made a 25,000-mile survey of the Near East and the lands bordering the Mediterranean. Lowdermilk, a professional in forestry and soil conservation with experience in China, wrote a report on his observations, *Conquest of the Land through 7,000 Years*, in which he maintained that deforestation and soil erosion had been a major factor in the decline of earlier civilizations in this region. This slender volume had great influence on American thinking about Mediterranean history and was not without influence in the Mediterranean itself. But it can also be read as a record of what he saw, as a keen and trained observer of the land and its vegetation, in the 1930s. In each country, he consulted agricultural experts, scientists, and officials to help him interpret the incomparable document of the landscape. In Mesopotamia, where cities and intensive agriculture originated, he visited the site of Babylon and

The impressive Roman-era ruins of Jerash, Jordan. Walter Clay Lowdermilk regarded the bared rock hills of the landscape around the city as evidence of the damage done by ancient civilizations to the land. (Photo courtesy of J. Donald Hughes)

found a landscape of "desolation," although in that alluvial terrain he judged that fertile soil was still in place. In Egypt he remarked upon the rising water table and salinization caused by the irrigation and continuous cropping made possible by the old Aswan Dam (to be discussed in the chapter on case studies). His travels took him across Sinai, where he saw evidence of accelerated erosion in recent times, and into Transjordan by way of Aqaba. Near Petra there were collapsed terrace walls and extreme erosion down to bedrock, with almost no vegetation. Crossing the Jordan, he noted vast areas of gullies and tracts stripped of soil. The denuded highlands of Judea were dotted with abandoned villages, in a land once flowing with milk and honey, a decadence he ascribed to poor soil management, although he also noted limited areas where methods of land conservation such as contour plowing and tree planting had been used, and he had a degree of optimism about the possible future restoration of the land. Back in Transjordan again, he described the bared rock of the hills around Jerash and admired the old waterwheels of the Orontes Valley in an area well cultivated but much reduced in population since ancient times. In northern Syria, a land of ghost cities, he thought erosion had done its worst. He marveled at Lebanon's "fantastic staircases" of terraces that existed even on mountain slopes with a grade of 76 percent, and lamented the loss of the great cedar forests, which were represented by fourteen small groves when he was there (he saw only four of them). He believed the trees could spread over Lebanon again if they were protected from the "rapacious goats that graze down every accessible living plant on these mountains" (Lowdermilk 1953, 13). Lowdermilk found that the island of Cyprus epitomized the land-use problems of the Mediterranean; there was a church in Asha that had been surrounded by 13 feet of alluvial silt since Byzantine times. In Tunisia he witnessed muddy flash floods pouring down the wadis during a winter rainstorm, an active illustration of the cause that produced many of the results he had been observing. Again, there were gullies and soil washed away to bedrock, with cultivation only in the valley floors where deposition had occurred. He found olive presses in places where there was "not a single olive tree within the circle of the horizon" (18), although in fact olives do grow in Tunisia where irrigation is possible. He doubted that the lack of olives was due to a change in the climate, since olive trees planted in modern experiments were flourishing. In Italy, Lowdermilk found a countryside that was densely inhabited, in contrast to the lands he had just visited across the Mediterranean. He praised the recent drainage of the Pontine Marshes and the attractive new farms there. The farmers of southern France also merited his approbation for conserving the soil in their terraces and practicing contour plowing. It is interesting to have such a glimpse into environmental conditions in the Mediterranean basin at this time when "environmentalism" had not yet found its contemporary meaning. But the idea of conservation was emerging, as well as a dawning of realization of the importance of environmental problems and the necessity to deal with them.

NATIONAL PARKS

An early emphasis of the environmental movement, in a time when its supporters were called conservationists, was the designation of areas of especial environmental interest, whether of outstanding natural features and scenery or the presence of rare or notable wildlife, and their protection and preservation. These areas were intended to retain their special character but were usually also open to visitors, whether local people or tourists. They received varying degrees of protection; sometimes designating an area as a national park meant very little, since few financial resources or none at all might have been available for staff or facilities. Some of them became quite popular and continue to be increasingly so. Visiting the Mediterranean for the purpose of "ecotourism" is now a widely recognized form of recreation. This aspect of conservation continues today, and every Mediterranean country has such reserves, including national parks, although in recent years the conflict between nature protection and the wear and tear caused by tourism has become acute in certain very popular parks that are, perhaps, being "loved to death."

Spain set aside two national parks in its northern mountains in 1918, and now has several national parks, including four in the Canary Islands. Portugal, by the way, has one also in Madeira where some of that island's original forest, the Laurisylva, survives. One of Spain's most interesting wild reserves is Coto Doñana National Park, an area of wetlands (*las marismas*) thronging with innumerable birds and other wildlife at the mouth of the Río Guadalquivir. Almost half of all bird species native to Europe have been observed there. It has sand dunes and pine forests as well, and is home to a population of lynxes.

The French Mediterranean coast is lined with a series of lagoons and wetlands in which flamingos breed. Camargue National Reserve is located nearby at the mouth of the Rhône; it is a place where nature and the life of the people reflect past centuries. There are white horses and black cattle, and French cowboys live much as their ancestors did. The nature reserve, an area of 85,000 hectares (328 square miles), was set aside in 1927.

Italy has a number of national parks; the first, Gran Paradiso in the Alps, was declared in 1919. Among the rest, the third, Monte Circeo, is well worth mentioning. Created in 1934, it preserves a wonderful bit of Mediterranean coastline in Latium, including a peak 541 meters (1,775 feet) high, thick forests and wildlife almost unimaginable in the region today, wetlands that represent what is left today of the famous (infamous?) Pontine Marshes, lakes, rocky shores, and an offshore island. A Neanderthal skull dating to about sixty-five thousand years ago was discovered in one of its caves. It is the mythic home of the enchantress Circe, who captivated Ulysses and turned his men into swine, and interestingly it was a famous place for boar-hunting in later antiquity. Theophrastus, the first Greek botanist, described its forests.

Greece has about ten national parks, although their protection has not been well organized over the years. The most famous is undoubtedly Mount Olympus, the supposed home of the gods, 2,911 meters (9,551 feet) in elevation and a very interesting botanical area. The park was created in 1938, along with another further south including Mount Parnassus, the mythical locale of Apollo, Dionysus, and the nine Muses.

Additional Mediterranean countries with national park systems are Turkey, Tunisia, Egypt, and Israel. The countries with the largest percentage of their land area within national parks in 1979 were Portugal, 6.5 percent, and Italy,

6.2 percent. Israel's Mount Carmel National Park was finalized in 1971. Libya has one national park. Some national parks are areas primarily of cultural rather than natural interest.

Other protected or partially protected areas, such as wildlife refuges and forest reserves, exist in most of the nations of the region. In the 1970s in Greece, for example, scientists recognized a forest area of mixed coniferous and deciduous trees that had escaped exploitation since time immemorial in the Rhodope Mountains near the Bulgarian frontier. This forest had been saved by its remoteness, difficult topography, and the fact that it is in a region where almost constant hostility had prevented development. Greece set aside a natural monument of 858 hectares (3.3 square miles) in 1975. Among the birds found there are capercaillie, golden eagle, and black and griffon vultures; mammals include bear, wolf, lynx, red and roe deer, and chamois. The scenery is exquisite, with mountains, gorges, streams, and waterfalls.

TECHNOLOGY

The great innovation of this period was mechanization, the use of metal machines driven by power generated by a heat source. The steam engine, first invented in ancient Alexandria but then used mainly as a toy or for special effects, was reinvented in the modern world, initially for pumping water out of mines, and subsequently found many uses, not least to power the railroads that revolutionized transportation throughout the Mediterranean. This produced changes in the traditional forms of transportation. For example, instead of being the major long-distance carriers, camels began to serve as haulers to the trunk railroads in Ottoman lands. The textile industry was one of the first to be mechanized; cotton-spinning mills replaced the spinning wheel. Ceramics and paper industries followed. Steam technology in sawmills increased their efficiency over the old water mills between eight and forty times, depending on the design, and would hasten deforestation proportionately. New techniques and equipment were devised for virtually every industry, including agriculture.

Mechanization of agriculture was difficult in the Mediterranean and therefore slow to appear. Dry summers and mountainous topography presented difficulties—the mountains made transportation high-priced, for example. Coal was less available than in the north, rendering energy expensive. The economic depression mentioned in Chapter 4 interfered with investment in the new machines, and a low level of literacy and education that persisted into the early twentieth century translated into a lack of skilled workers. Italy lacked coal resources and had to import that material, but in most other ways it outstripped

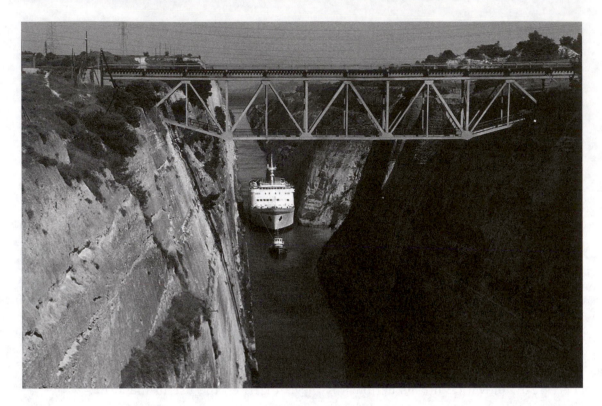

The canal through the Isthmus of Corinth saves ships the longer and sometimes perilous journey around the southern capes of Greece. It was completed in 1893, and the deepest cutting is 87 meters (285 feet). The Roman emperor Nero ordered its construction in the first century AD, but Roman technology was inadequate to the task. (Photo courtesy of J. Donald Hughes)

many of the other Mediterranean countries and moved closer to the pattern of central and northern Europe, especially in Milan and its region. Catalonia, the linguistic province around Barcelona, became industrialized more rapidly than the rest of Spain. Most backward of all in this respect were the countries that had been under Ottoman occupation for some four hundred years, where governments were inefficient and finances chaotic, and where the former regime had made a custom of soaking up all agricultural surpluses. Greece, for example, had little iron or coal, an inadequate system of transportation, and limited markets. In 1913 its manufacturing production per capita was 9 percent of Britain's, while Italy's was 23 percent of Britain's.

Steamships greatly reduced waterborne travel times both on the sea and on navigable rivers such as the Nile and Tigris, where the improvement of upstream speeds was close to miraculous. On average they could carry from ten to twenty times the tonnage of sailing ships. The opening of the French-built Suez Canal in 1869 created a second entrance (and exit) to the Mediterranean, greatly reducing cost and time on many sea routes, and providing an exchange of species of fish and aquatic organisms between the Mediterranean and Red seas.

INDUSTRIES

In the early days of the Industrial Revolution, energy for steam was provided by wood and charcoal, which were demanded in ever-increasing amounts. The effects of these demands seemed to make the destruction of the remaining forests imminent, and certain governments became concerned. The centralized government of France was able to initiate protection in the Mediterranean districts under its control, but Italy was fragmented into a number of jurisdictions at the time. The other Mediterranean countries were only marginally industrialized. The shift to coal at the beginning of the modern era gave the forests some respite, but certainly not exemption, from loss. Many industries moved from forested districts to areas where there were coalfields. Italy could not make this aspect of the transition because it possessed no important coal deposits. The human environment suffered from greatly increased pollution; wood smoke had been bad enough, but the thicker, blacker coal smoke, full of particulates and laced with sulfur dioxide and other damaging chemicals, was far worse. When the internal combustion engine came into use, burning petroleum products, its emissions added various organic and nitrogen compounds to the mix, where photochemical reactions took place driven by the energy of sunlight, which is noted for its strength in the famously sunny Mediterranean. All the large cities of the region were at times blanketed by smog, from Madrid and Rome to Athens and Cairo. Coal smoke tended to come from sources such as power plants and factories at fixed locations, or from trains that traveled tracks on regulated schedules, and these were in theory more easily controlled; laws were sometimes passed to do this, but the industries and their leaders were wealthy and politically powerful and often managed to avoid regulation. The products of petroleum combustion came mainly from cars and trucks, and both the ownership of these by the general public and the means to control their emissions occurred only very late in this period.

MODERN EGYPTIAN ENVIRONMENTAL HISTORY

Egypt in the years just before 1800 was an impoverished, cutoff, and ignored province of the Ottoman Empire. But spices from the east and gold from the south still moved through Suez and Alexandria, even if in considerably smaller quantities than before. Egypt continued to produce sugar, rice, and wheat; to spin linen, cotton, and woolen thread; and from them to weave textiles suitable for export. It was, then, not its resources and economic importance but its strategic location in respect to British interests in the Mediterranean and India that led Napoleon to launch a French invasion of Egypt in 1798. He evaded the British fleet, captured Malta en route, and landed at Alexandria. Napoleon's military force lost more men from mosquito-borne diseases, dysentery, and thirst than from his Mamluk and British adversaries. The environment was his greatest enemy. At the same time, he did study it, since he brought with him some 167 scientists and scholars whose labors produced the multivolume *Description de l'Egypte*, which catalogued not only the archaeological treasures of the country but also its biological, agricultural, and mineral wealth.

Although successful against the Ottomans, Napoleon eventually lost to the British in 1801 and abandoned Egypt, but his occupation had set in motion a process of economic and political change. An Albanian officer in the Ottoman army in Egypt, Muhammad Ali, seized power in 1805 and began the modernization of Egypt. Under his rule, the state gained control over the land, the water, and the development of agriculture. With aid from French engineers, he established an irrigation system and a transportation infrastructure that made possible the production of three harvests a year on much of the land that had only produced one, so that the land and the peasants would be at work continuously raising crops for export and for a positive balance of trade. The peasants were told what to plant and when to plant it, and though they were paid better than before for the crops they raised, they had to work 250 days in the year instead of the 150 they had been used to, and in addition had to perform paid corvée labor on the irrigation systems. The crops were for the most part not food plants but cash crops, including long-staple cotton, sugar cane, tobacco, and indigo (a raw material for dyes).

Ali introduced hundreds of varieties of plants, trees, and fruits. He attempted to start new enterprises by bringing in mulberry trees and silkworms from Syria and Lebanon to begin the production of silk, and cashmere goats from north India to serve as the basis for a fine wool industry. Both of the ventures just named were financial failures, since the organisms involved did not flourish in the Egyptian climate. More successfully, he encouraged the opening of factories under foreign managers to manufacture textiles from domestic materials, paper, soap, ships, gunpowder, and arms including rifles, muskets, and cannon. Western-style education, including industrial schools, appeared alongside the Islamic madrasas that had long been famous in Egypt.

Muhammad Ali's eldest son, Sa'id, became Ottoman viceroy of Egypt in 1854 and granted a French company under the former diplomat Ferdinand de

The Mosque of Muhammad Ali on the citadel of Cairo, Egypt, was built in part of stone taken from the pyramids of Giza. The dome was constructed using cedar timbers imported from Lebanon. Muhammad Ali, who ordered its construction, was ruler of Egypt from 1805 to 1854. (Photo courtesy of J. Donald Hughes)

Lesseps permission to construct a saltwater canal across the Isthmus of Suez between the Mediterranean Sea and the Red Sea. This was an attractive idea to him because it would facilitate the movement of Egyptian exports and provide a source of income from foreign shipping. The canal was inaugurated in 1869. It is entirely at sea level, and thus no locks served as barriers of any kind. By making the long voyage around Africa by way of the Cape of Good Hope unnecessary, it greatly increased the amount of shipping in both directions through the length of the Mediterranean Sea. It also allowed the passage of aquatic organisms between the Red Sea and Mediterranean Sea, either directly through the canal or by being carried in ballast water on board ships. At least forty-one species of Red Sea fishes, and an unknown number of crustaceans, mollusks, and other marine invertebrates, have become established in the Mediterranean as a result of the canal. Some of them have spread throughout the eastern Mediterranean basin. There has been less movement in the other direction, apparently because Mediterranean species find the temperature and salinity of the Red Sea less hospitable.

Egyptian exports were accelerated by the U.S. Civil War, because the federal blockade of Confederate seaports cut off Europe's major source of cotton. As a result Egypt experienced a cotton boom until American cotton exports gradually resumed during the 1870s. The resultant decline in exports may not have been an unmitigated disaster, since cotton's ascendancy had taken up land that could have been used for food production, and repeated plantings of cotton in the same area had depleted the soil.

Meanwhile, Egypt's balance of trade had faltered and its debts had risen. British forces occupied Egypt in 1882, effectively severing Egypt from the Ottoman Empire and giving Britain control of the Suez Canal. A British financial adviser was sent to organize an Egyptian Debt Commission, which would enact measures to avoid bankruptcy, and in the following year Britain imposed Sir Evelyn Baring, Lord Cromer, as de facto governor. The army, bureaucracy, and schools were cut back, but the irrigation infrastructure was repaired and improved in order to boost agricultural production and tax revenue. The most notable water project was, of course, the first Aswan Dam, which is discussed in the third case study in this book.

Egyptian resentment against continued British control and resultant political agitation managed to win British recognition of Egypt as a sovereign state, with some reservations, in 1922. Environmentally, Egypt in the 1920s and 1930s remained an agricultural land dependent on long-staple cotton exports. The area under cultivation enlarged considerably, but the population that had to live on the land increased even more rapidly, and the segment of the population that increased to the greatest extent was composed of poor landless peasants. In addition to competition from the new artificial fiber industry abroad, cotton production suffered from ecological problems such as cotton maggot infestation and dropping yield per hectare due to saturation from continuous irrigation and ensuing salinization of the soil. Then came the worldwide depression and a disastrous drop in cotton prices, so that small landowners went bankrupt, could not pay taxes, and lost their land. Many of the poor moved to the cities to seek jobs there. In rural areas, bilharzia and trachoma were endemic because there was almost no piped drinking water or sanitary sewage disposal, with the result that the people used water from polluted ditches and ponds. Water buffaloes and chickens shared living quarters with the peasants.

During World War II (1936–1945) staple foods such as wheat, corn, and rice turned into expensive scarcities, and much of the little that was available was requisitioned by British and Allied armies (and would have been commandeered by German troops had General Erwin Rommel's invasion of Egypt not been stopped narrowly at the battle of El Alamein fifty miles west of Alexandria). Agricultural landowners and manufacturers made a profit on food and on producing scarce items such as cigarettes, alcohol, and sugar.

Many factors led to the nationalist revolution of 1952 that brought Gamal Abdel Nasser to power, including the Egyptian defeat in the Palestine War of 1948–1949, the corrupt government of King Farouk, and continued British influence. But a major cause was also environmental and economic. There was an agricultural crisis, and the landowning class had become an intolerable burden for the common laborers in the countryside, while population growth and flight to cities had engendered an urban underclass of unemployed youth who saw hope in political action. Nasser's government soon took steps toward land reform and toward job creation by education, industrialization, and the construction of a new Aswan High Dam. (The story of the latter project is told in the case study on the dams at Aswan.)

DISEASE AND ENVIRONMENT

In the course of the modern period, especially after 1850, the advance of medical science, the organization and availability of medical services, and environmental health and sanitation measures began to make major inroads against epidemic diseases, producing higher survival rates and allowing accelerated population growth.

One of the diseases most amenable to the improvement of environmental conditions, especially the provision of a dependable safe water supply, was cholera. The microorganism that causes this rapid and often fatal disease is carried in water polluted by human bodily wastes, a fact that was unknown until

John Snow demonstrated it circumstantially in London in 1854. Cholera was endemic to India, especially to the Ganges Delta, where conditions for its spread prevailed. The geographic spread of cholera was slow because its victims could rarely travel far before they died, but this circumstance changed with the development of more rapid means of transportation in the nineteenth century. The first outbreak to reach the Mediterranean area occurred in 1821, when Muslim pilgrims from India infected others in the Arabian port of Muscat, and returning pilgrims carried it to Syria and Palestine. The severe winter of 1823–1824 appears to have snuffed out this epidemic. But in 1831, pilgrims conveyed cholera to Mesopotamia, Syria, Palestine, Egypt, Tunis, and beyond. In Cairo, 13 percent of the population perished, and during the crisis the ruler, Muhammad Ali, asked a number of European consuls to constitute themselves a Board of Health for the city, which they did.

The risk of contracting cholera on the journey to and from Mecca did not dissuade many Muslims from undertaking one of the five basic requirements of the faith. A pilgrim had to accept danger as part of the effort, and to die while on pilgrimage was considered to be a blessing. Between 1831 and 1912, cholera appeared forty times during the period of pilgrimage. The Mediterranean basin experienced it almost as often. Cholera came to Portugal and Spain from England in 1833 and spread from Barcelona to Marseille in the following year. It visited Malta in 1837. There was a third pandemic in 1846 and a fourth in 1863. When cholera returned to Egypt in 1883, European doctors' advice was sought again, and they advised the use of chemical disinfectants, with some positive effect. In the same year Robert Koch identified the bacillus, and ten years later a vaccine against cholera was developed, although it was only partially effective. Even so, Egypt legislated compulsory inoculation against cholera in 1913. Cholera has not disappeared from the Mediterranean or the world today, but it can be medically treated, and its prevention through safe, treated water supply is understood. Public water systems have been established in many Mediterranean cities. In Athens, for example, the Marathon Dam and treatment system were constructed in the 1920s, and they continue to supply clean water to the city. There, however, and in many of the rapidly expanding cities of the southern and eastern Mediterranean, provision of freshwater has not been able to keep up with the growing urban populations.

After it was demonstrated that the mosquito is the vector of malaria, environmental control of the disease seemed possible. Even as late as the 1880s, however, citizens of Ravenna went to bathe in the sea on August 10, the feast of San Lorenzo, in the belief that the saint would protect them against malarial fever, even though the beach was next to mosquito-breeding freshwater swamps where the insects swarmed in summer. Mosquito control was used against malaria in Greece and other countries in the 1920s; usually this involved getting rid of standing water in and near human residences, and covering mosquito-breeding ponds with a thin layer of oil. But during and after World War II, the insecticide DDT became available and was applied throughout the environment—both outdoors in every imaginable place and indoors on the interior walls of dwellings where mosquitoes rested. Malaria was eradicated from many districts, and it seemed to some enthusiastic observers that it had been defeated permanently. But ecological contradictions soon appeared. DDT killed beneficial insects as well as mosquitoes, and it built up in the tissues of birds and other predators of mosquitoes, causing reproductive failures and even deaths. Meanwhile, the numbers of mosquitoes began to recover because the portion of the mosquito population that had genetic resistance to DDT survived and multiplied. Humans inadvertently had created the conditions for the ecological evolution of mosquitoes that are resistant to DDT. Other pesticides, some of them with effects as bad as or worse than DDT, followed in a process of diminishing returns. Malaria remains a danger in more ways than one, since the malaria parasite itself has also developed resistance to some of the medicines commonly used against it.

Typhus is transmitted by lice, which can be controlled by personal and public hygiene, but both flourish in conditions where such hygiene is difficult, such as in armies in the field. At the beginning of World War I, Austria delayed the invasion of Serbia due to reports of an outbreak of typhus there, and the delay affected the course of the war. In the terrible conditions of World War II, typhus infested the civilian population of Naples. Here the American military occupation undertook a wholesale compulsory delousing of the city population, which involved dusting with DDT under their clothing. This time the use of the pesticide was successful, and it was the first time in history that a typhus epidemic of this size had been brought under control.

The eradication of smallpox was another achievement of public medicine during the modern period. The cowpox vaccine (described in Chapter 4) was improved and made widely available, so that vaccination against smallpox was made compulsory in several countries between 1880 and 1900, including Serbia, Austria, and Egypt.

Poliomyelitis appeared in Italy in 1883. It was prevalent throughout the world until 1954, when an effective vaccine was developed, and used widely in subsequent years, so that the disease became very rare. It seems that this form of prevention is more effective, and ecologically more benign, than the large-scale broadcasting of biocides.

CONCLUSION

In the nineteenth century and the first two-thirds of the twentieth century, the relationship of humans to the natural environment in the Mediterranean region changed significantly. Part of this change consists of the fact that the power of human technology to affect the natural environment greatly increased. This is not to say that in the previous centuries humans lacked the upper hand; indeed, the Mediterranean region had been subject to the modifications caused by civilizations as long as, or longer than, any other region on Earth. It is, rather, a matter of scale and speed: of the energy available to humans, the sophistication of technology, the sheer volume of the economic movement of materials, the extent and variety of pollutants generated by human activities, the increase of population beyond any level seen before, and the accelerating pace of all these changes.

The damage to natural systems was consequently greater than that of previous times. It may seem that the story of environmental deterioration has repeated itself in each of the preceding historical chapters, with the loss of soil and forests, the decline of wildlife on the land and in the sea, and the exhaustion of nonrenewable resources. But the severity and geographical extent of the damage increased by orders of magnitude during this period, and would continue to do so exponentially in almost every respect afterwards.

Before the modern era, and well beyond its beginning, the almost universal attitude toward human changes in the environment had been that they were improvements. The Roman orator Cicero in the first century BC had praised the cleverness of human hands, describing the many things they can accomplish, including agriculture, forestry, navigation, and hydrology, and concluded approvingly, "Finally, by means of our hands we endeavor to create as it were a second world within the world of nature" (*On the Nature of the Gods* 2.13). Later, Christians gathered from the Hebrew Bible that God had designed the Earth as a garden of everything good, intended for his human children "to till it and to keep it" (Genesis 2:15). The Qur'an states, "It is He who has spread out the Earth for his creatures: therein is fruit and date palms, producing dates, also corn with its leaves, and stalk for fodder and sweet smelling plants. Then which of the favors of our Lord will you deny?" (Qur'an 55:1–13). This image of beneficent use was dominant for centuries; a countryside full of farmhouses, fields, and domestic flocks was considered superior to the forsaken wilderness of howling forest and desert, and the work of human beings was considered to be inevitably an improvement. In the modern era, however, it became increasingly evident that the human reshaping of the landscape was not always good. In a conclusion based largely on his observation of the Mediterranean area, George Perkins Marsh judged,

But man is everywhere a disturbing agent. Wherever he plants his foot, the harmonies of nature are turned to discords. The proportions and accommodations which insured the stability of existing arrangements are overthrown. Indigenous vegetable and animal species are extirpated, and supplanted by others of foreign origin, spontaneous production is forbidden or restricted, and the face of the Earth is either laid bare or covered with a new and reluctant growth of vegetable forms, and with alien tribes of animal life. (Marsh 1965, 36)

Reflecting on the idea that human actions could have destructive consequences, there were intellects during this period who began to urge the preservation and restoration of the environment. Governments enacted laws protecting certain species, and they set aside certain areas as national parks and wildlife reserves. The effectiveness of such conservation regulations depends on their enforcement, and it must unfortunately be said that they were more evident on paper than they were in the field. Policing the forest and the sea is expensive, and funds are not always available. Even so, the intent was there, and as public awareness began to grow at the end of the period, it was possible to hope for improvement.

References

Carrington, Richard. 1971. *The Mediterranean: Cradle of Western Culture.* New York: Viking Press.

Lowdermilk, Walter C. 1953. *Conquest of the Land through 7,000 Years.* U.S. Department of Agriculture, Soil Conservation Service, Agriculture Information Bulletin No. 99.

Lowenthal, David. 2000. *George Perkins Marsh: Prophet of Conservation.* Seattle: University of Washington Press.

Marsh, George Perkins. 1864, reprinted 1965. *Man and Nature.* Ed. David Lowenthal. Cambridge, MA: Harvard University Press.

6

CONTEMPORARY TRENDS AND CONCERNS (1960–2004)

A dangerous complacency has crept in and people are too ready to talk about the sea having been saved, while ignoring the almost universal failure, time and time again, to carry through the most basic of changes. Real change will come only when those in power have become convinced that environmental protection is necessary and urgent. Without it, the Mediterranean will continue its slide into chaos.

—XAVIER PASTOR, *THE MEDITERRANEAN*
(*GREENPEACE: THE SEAS OF EUROPE*)

The Mediterranean is a representative microcosm of the entire Earth.

—MOSTAFA KAMAL TOLBA, EXECUTIVE DIRECTOR,
UNITED NATIONS ENVIRONMENT PROGRAMME,
IN *FUTURES FOR THE MEDITERRANEAN BASIN*

The contemporary period begins with the years of economic growth and recovery after World War II. This was a time of unprecedented growth in population, technology, and industry, but also in public awareness of pollution and concern about other environmental issues. The years from the end of the twentieth century through the beginning of the twenty-first century demonstrated that, as John R. McNeill put it, "The human race, without intending anything of the sort, has undertaken a gigantic uncontrolled experiment on the Earth" (2000, 4). As he also noted, the uniqueness of this period has to do not so much with new kinds of environmental change as with the scale, intensity, and speed of change. Considering the entire stretch of human history, the Mediterranean area was one of the first sections of the globe to feel human-induced environmental impacts. In this respect, probably the only other part of the Earth that can be compared with the Mediterranean is China, although India comes close. Ecologically, it bears the scars of long occupation by human civilizations, and more recently it has for the most part joined the rapid use of resources and technological development characteristic of the globe now dominated by a world market economy.

The political pattern that emerged throughout the Mediterranean region after World War II is unlike any that existed there before. A series of nation-states, some centuries old and others quite new, rings the basin. Some of them are historically hostile to one another, and some are in conflict, or have been recently. The most notable conflict of the past half-century has been that between the state of Israel, created in 1948, and its Arab neighbors including Palestine, which has not yet received full recognition as a state. Thus the idea of international cooperation for the environmental protection and restoration of the Mediterranean seems extraordinarily problematic. Yet, as shall be noted in this chapter, some important advances in cooperation have been made.

A number of important political changes occurred after World War II. The northern European empires relaxed their grip on Mediterranean lands, in some places quickly and in others reluctantly. France granted Algeria, its last North African colony, independence in 1962. The United Kingdom released Cyprus in 1960 and Malta in 1964, but still hangs on to Gibraltar. Spain keeps two small enclaves on the Moroccan coast, Ceuta and Melilla. Turkey invaded and occupied northern Cyprus in 1974, setting up a government of the Turkish minority on the island that has not been recognized by any other nation. The 1990s witnessed the fragmentation of two important political federations, the Soviet Union and Yugoslavia. Slovenia, Croatia, and Bosnia severed their ties with Serbia in 1991, and Macedonia in 1992, beginning a series of destructive conflicts that culminated in a NATO intervention in Kosovo, a province of Serbia, in 1999. Yugoslavia was left with two constituent republics, Serbia and Montenegro, and Montenegro also appears to be moving toward independence. Ukraine and Georgia withdrew from the Soviet Union in 1991 and now share the coasts of the Black Sea with Russia, Turkey, Bulgaria, and Romania. At present, therefore, the Mediterranean Sea is bordered by about twenty independent, self-governing nation-states that are members of the United Nations, and two political entities, Palestine and North Cyprus (the area invaded by Turkey in 1974), with limited international recognition.

A number of environmental problems particularly affected the Mediterranean area in the last third of the twentieth century. Among these is water pollution, both in the rivers and in the sea, which is severe because the Mediterranean is almost landlocked. Materials that enter it, whether sewage or

The popular beach of Sitges south of Barcelona, Spain, illustrates the crowding and development that mark the impact of tourism on the Mediterranean coasts. (Photo courtesy of J. Donald Hughes)

chemical wastes, remain within it and accumulate over time. Then there is the problem of water supply, due to the generally low rainfall in the Mediterranean basin. The use of water by various segments of the economy varies greatly around the basin; in Greece, 80 percent of water demand is represented by agriculture, whereas in France the demand from industry, including hydroelectric use, which in France is mainly water used to cool nuclear power–generating facilities, amounts to almost 75 percent of the total. How to divide the water available in rivers such as the Jordan, Euphrates, and Tigris is a question that must be addressed in any peaceful solution for the troublesome conflicts of the Near East. Possible desertification of Mediterranean lands has been the subject of an active research program of the European Union. Another problem is overdevelopment of the coastline with industries and tourist facilities including hotels, with attendant erosion and loss of littoral habitats, an unfortunate result of the inherent attraction of the Mediterranean's mild climate and scenic beauty in which tourism is destroying the very reasons for its existence. Atmospheric pollution is unfortunately very noticeable in this region: the air is often

This view of Athens, showing Mount Lycabettus, illustrates the facts of urban sprawl and pollution in the contemporary Mediterranean area. (Charles O'Rear/Corbis)

stagnant and subject to temperature inversions, as anyone who has tried to drive in Athens during the chronic *nefos* (smog) is aware. In the 1970s, Athens had an average nitrous oxide pollution level twice that of Los Angeles, and a particulate pollution level 47 percent higher than Los Angeles. When Athens, in an attempt to reduce air pollution, passed a law excluding cars from the city on alternate days, based on whether the last digit of the license number was odd or even, the response of many affluent citizens was to buy a second car and to insist on a license plate that bore the opposite kind of number, thus enabling them to travel into the city every day.

The reduction and loss of native animal and plant populations are troubling, both on land and in the sea. Some of this is due to pollution, and some is due to encroachment of other forms of land use into wildlife ecosystems. Indiscriminate hunting of every sort of creature and overfishing present threats to biodiversity. There are massive invasions of exotic species that crowd out the indigenous Mediterranean sea life. Wars have been damaging to the environment. Problems related to energy development, including use of fuel resources and nuclear power plant accidents and releases of radioactivity to the environment, have presented disturbing challenges to the responsible businesses and agencies. Fleets with nuclear weapons prowl the sea, and nuclear bombs have inadvertently fallen into its depths, fortunately without detonating. Among the Mediterranean nations, France and Israel possess nuclear weapons. In addition, Mediterranean cities do not ordinarily lack any of the urban problems that plague cities around the world, with the possible exception of snow removal, although that, too, has been needed at rare times even in Athens and Jerusalem.

POPULATION AND HUMAN SETTLEMENTS

The immense and rapid increase in population generally, and urban population in particular, is one of the most important forces in the human relationship to the natural environment, increasing human impact and deteriorating the environment. In addition, the Mediterranean basin emerged as one of the world's leading tourist destinations in the years after World War II. Today there are about 160 million residents near the Mediterranean seacoasts, and annually more than the same number of foreign tourists spend some time there. In 1980, for example, Italy alone had 66.2 million visitor-nights of foreign tourists in its hotels, and Spain had 58.7 million. Tourism's explosive growth has been particularly noticeable in the Mediterranean islands, since their maritime geography makes them attractive to tourists, and their small size places space for tourist facilities at a premium. Tourism is the leading source of foreign exchange for the island states of Malta and Cyprus, and the islands of the Greek archipelago and the Balearic Islands of Spain draw a large proportion of each country's foreign visitors. Almost all the Mediterranean nations have made successful efforts to increase tourism, although Libya until recently chose not to do so for politico-religious reasons, while other states suffer from instability and lack of infrastructure. Since tourists in the Mediterranean area tend to be of the traditional touristic type that seeks beaches and sun, development crowds the coastline, destroying the environmental amenities of easy access to the marine environment and the littoral ecosystems. In countries with little control on coastal development, such as Greece and Morocco, large numbers of tourists may actually be excluded from the beaches they came to visit by the solid ranks of tourist facilities that now form barriers there. Even though tourism's positive inputs to governmental finances are undeniable, the pressures brought to bear by tourism on the coastal milieu are far from harmless.

Cities increased their human numbers more rapidly than the countryside and came to include a much greater percentage of the total population in every country. This was certainly true of the Mediterranean, but it was also true of the world at large. The most considerable increases took place in developing countries. Cairo, the largest city in the Mediterranean area, grew from 1.3 million in 1937 to 10 million in 1997, placing it among the five most populous cities in the world. It is predicted that Cairo will triple in population during the twenty-first century. Other large Mediterranean cities in the late 1990s were Istanbul (8.3 million), Baghdad (3.8 million), Alexandria (3.4 million), Athens (3.0 million), Ankara (2.9 million), Casablanca (2.9 million), Madrid (2.9 million), and Rome (2.7 million). Others with between 1 and 2 million included, in rough descending order of size, Milan, Barcelona, Aleppo (Syria), Algiers, Damascus, Naples, and Rabat. Athens doubled in size between 1950 and 1970, while most of the Greek countryside was losing population and was far from atypical in this picture of urban growth and rural loss. In 1990, about half of the population in Mediterranean countries lived in urban areas: 69 percent in the northern basin and 47 percent in the south.

Urbanization affects people and environment not just in cities, but also far beyond the city limits, since cities depend on the resources of the countryside in their immediate neighborhoods and also the resources of more distant areas with which they trade. The environmental damage, or "footprint," of a city may extend to lands hundreds of kilometers or half a world away. Cities demand adequate supplies of water, fuels, and other natural resources from near and far. In return they offer manufactures and trade. The metabolism of cities also excretes pollution, sewage, garbage, and other wastes, which severely deteriorate the environment when they are allowed to remain within cities, but the more the city is able to improve the living conditions for its inhabitants, the more of these embodiments of entropy have to find a sink somewhere else in the environment. About 40 percent of Mediterranean cities lack any kind of sewage treatment facilities.

Cairo certainly represents virtually all of the environmental problems of the contemporary Third World city. Housing is high among these; the most-used material, brick, is for the most part baked mud taken from the soil of the Nile Valley, meaning that each new building decreases the agricultural base. A housing shortage, along with the poverty of a major segment of the people, results in the fact that many families live in shanties among the tombs next to the garbage dumps from which they make part of their living. The water supply comes almost entirely from the Nile River, which is polluted by sewage and agricultural chemicals. Those who cannot afford filters or bottled water face illness, perhaps even death, from drinking tap water. But less than half of Cairo's population even has tap water, and those who do have it find that it runs only part of the day and is subject to long cutoffs. Garbage disposal is a problem recognized by all the residents; it is often disposed of carelessly outside the containers that are provided in some areas of the city, and is not systematically collected. As a Cairo taxi driver told an Egyptian graduate student researcher, "We could live with the water problem, especially because we are prepared for the long cut offs, yet what is really nagging us is the garbage and sewage in the streets. These are the real problems that cause ill health, especially because we can do nothing to solve them. It is a complete disaster when you get out and find the streets in this filthy condition" (Gomaa 1995, 10). Despite strict laws on the books, hazardous waste materials are disposed of in dangerous ways. A number of air pollutants grossly exceed international standards. The sources of these include dust churned out by cement plants, motor vehicle emissions from leaded gasoline in poorly maintained engines moving (or not moving) in slow traffic, and factory smoke. Incineration of wastes often compounds the problem; gases emitted by burning plastic and other materials are often toxic. A natural pollutant—the sand blown in by the spring winds, the *hamsin*—gets into everything including the food, constituting a problem more specific to Cairo and cities in Upper Egypt. Cairo is perhaps an extreme case, but similar problems are found in other urban concentrations on the southern and eastern margins of the Mediterranean.

Cities in the European section of the Mediterranean basin have not grown as rapidly as those in African and Asian sections, and their growth began sooner and took longer, so that they have had the chance to create more rational plans and to build a better infrastructure. Streets were straightened and paved, drainage and sewage treatment were improved, safe water supplies were ensured, and public health measures increased life expectancy and reached a point where cities experienced a natural increase of population instead of depending on immigration to make up their losses. But these improvements were only a matter of degree, and serious problems remain.

Acid rain caused by the combination of acids of sulfur and nitrogen with water vapor in the atmosphere reacts with limestone and other materials used in public buildings and works of art, many of them very historic. The sculptures and buildings of Venice and the Acropolis of Athens are examples. In the case of Athens, many statues and reliefs have been removed from the Parthenon and other ancient monuments and placed in museums, sometimes inside chambers filled with inert gases, their places on the buildings occupied by replicas in materials whose colors are not entirely convincing. In Rome, the gilded bronze statue of the emperor Marcus Aurelius now stands in a room behind glazing to protect it from the polluted air and rain, instead of the center of the piazza on the Capitoline Hill that was its majestic former location.

Venice, a remarkable historical survival of a beautiful medieval city supported on wooden pilings driven into the marshlands at the northern end of the

Canals serve as streets in Venice, Italy, but with subsidence and rising sea level, tides and storms often bring high water that invades the lower stories of buildings like these, and erodes their masonry. Venice is considering measures to counter these dangers. (Photo courtesy of J. Donald Hughes)

Adriatic Sea, seemed toward the end of the twentieth century to be sinking back into the sea. Venice is located on islands in a lagoon about 52 kilometers long, which is in turn protected from the Adriatic Sea by a row of long, sandy islands. The tides flow in and out through narrow inlets between the barrier islands twice a day, bringing in fresh seawater and carrying out water polluted by the effluents of the city and the industries surrounding it. In November 1966 a high tide driven by sirocco winds from North Africa forced water into the city at 2 meters (6.3 feet) above average sea level in the almost tideless Mediterranean. The piazza of St. Mark's became part of the sea for fifteen hours, and the ground floors of buildings throughout the city were filled with water. Floods had invaded the city before, but never so high. The 1966 flood, however, was only a harbinger of others to come. By 1989, St. Mark's Square was covered by water forty times a year, and by 1996, almost one hundred times. The Doges' Palace and St. Mark's Basilica had to be protected by sandbags. Of course the rest of the city also suffered, and between 1950 and 2000 the historic city lost two-thirds of its residents. This is understandable because the effects of high water in the city are very unpleasant. Cisterns used to collect rainwater for drinking were invaded by seawater and became brackish. Furniture and stored foods and other materials were ruined. In churches, the water entered tombs in the crypts and accelerated the putrefaction of corpses, resulting in disgusting odors that made the buildings unusable for months afterwards. When the water rose above the protective stone foundations and reached the brick and plaster above them, the walls began to deteriorate rapidly due to cycles of salt crystallizing and dissolving. Floods almost equal to the 1966 level occurred in 1996 and 2000. The cause of the increasing floods has been blamed on the pumping of groundwater from beneath the city and its surroundings and thus lowering the ground level, the filling of parts of the lagoon for industrial purposes and leaving less area for the incoming water caused by the tides to spread out, and the general rise of sea level due to global warming. It seems quite certain that if nothing is done to ameliorate the processes of destruction, Venice will be lost to the sea within the next fifty years.

Several plans have been proposed to save the city, but the one most seriously considered is to position a series of seventy-nine huge steel gates on the seafloor between the islands, which in the case of threateningly high tides would be filled with air and raised high enough to block the sea from entering the lagoon. Locks would be constructed to allow ships to pass around the gates. Opposition to the project has arisen on the grounds that the gates would also prevent much of the outflow, rendering the lagoon yet more stagnant and sewage-filled. In early 2003, after long controversy, the Italian government granted appropriations to complete the design of the project and move toward its construction.

In speaking of human settlements, the creation of Israel, recognized by the United Nations in 1948, can be noted as one of the most significant transfers of population in the twentieth century. Many of the Zionists who advocated Jewish settlement had an environmental perspective, whether it was voiced in religious or secular terms. They advocated a new, positive relationship to nature along with a return to a land that would be valued and restored. The tree-planting programs and the establishment of agricultural kibbutzim where Jews could reestablish their connection with the earth are examples of this. The contemporary problems of the Middle East have an environmental basis, since the conflicts accompanying the founding and expansion of Israel uprooted many Palestinian Arabs who lost land with which they had a deep historical relationship.

POLLUTION OF THE MEDITERRANEAN SEA

If any chemical waste, sewage, or other pollutant enters the Mediterranean Sea, the only way it can leave is by following the outflow that exits below the much larger inflow of seawater at the Strait of Gibraltar. Theoretically, the Mediterranean would make a complete exchange of its waters with the Atlantic in 80 years, and the Black Sea in 140 years (although the deepest Black Sea waters are isolated from the surface waters and probably never are renewed; they are therefore anoxic). Obviously the outflow to the Atlantic is a minuscule proportion that has little effect on the pollutants in the western Mediterranean, which flow in more rapidly than they are removed, and virtually no effect on those in the eastern Mediterranean or Black Sea. The unfortunate situation, therefore, is that the Mediterranean Sea is a sink for the effluents of modern industry, including polychlorinated biphenyls (PCBs); the outwash of fertilizers, pesticides, and manure from industrialized agricultural lands; every imaginable variety of plastics and packing materials; and human wastes. Unfortunately, industrial wastes and urban wastewater are poured directly into the sea at many points around the coasts, and pollutants from inland reach the sea by way of rivers. Ships discharge pollution as they move; oil tankers are particularly damaging in this respect, due to routine releases of ballast water as well as accidental oil spills that are all too common. It should be remembered that the Suez Canal is at the Mediterranean's eastern extremity, so that about one-quarter of the world's oil shipping crosses the Mediterranean from end to end. In addition, offshore oil production from platforms in the sea is a significant source of pollution. The Mediterranean receives more than 600,000 tons of oil pollution on average each year, making it one of the most oil-polluted seas in the world, and its beaches spotted with tar reflect that fact. Tour vessels also dump garbage and sewage, and fishing boats jettison unwanted dead fish and every kind of trash imaginable. The total amount of pollution that pours into the Mediterranean Sea every year (including inflow from rivers) now includes approximately 500 million tons of sewage, 820,000 tons of petroleum, 600,000 tons of nitrate fertilizer, 200,000 tons of phosphate fertilizer, 60,000 tons of detergents, 25,000 tons of zinc, 4,800 tons of lead, 2,800 tons of chromium, and 100 tons of mercury (Harrison and Pearce 2000, 147; Attenborough 1987, 206–207). As a result, much of the seawater in the Mediterranean is laden with microbe-rich sewage, fertilizers and other nutrients, and synthetic organic chemicals such as PCBs and pesticides like DDT (outlawed in Europe but still used in the less developed countries), along with heavy metals including but not limited to lead, chromium, zinc, and mercury, and radioactive materials. Fish in the Adriatic Sea have been shown to have unusually high levels of mercury, and local people who eat them have sometimes accumulated toxic levels of that metal. Seabirds have suffered breeding failures or death. "Red tides" of toxic algae, blooms caused by eutrophication, became increasingly common from the late 1960s onward, killing many thousands of fish, crustaceans, and mollusks. The danger of

Waste dump on the beach at Maddalena National Park, Sardinia, Italy. Many beaches around the Mediterranean are degraded by dumps like this one, or by trash washed up by waves. (ML Sinibaldi/Corbis)

the brew, in particular the microorganisms it contains, to those who bathe in the sea or eat seafood is evident; dysentery and other gastrointestinal problems are common and sometimes fatal, along with hepatitis, typhoid, and other diseases. Many species of seabirds have been observed to ingest pieces of plastic, which they sometimes feed to their young. Others become entangled in larger pieces of plastic such as those that hold six-packs of cans together. The most polluted areas are in the north—especially the Adriatic Sea and the coastal waters of northeastern Spain—southern France, and northwestern Italy. However, no section of the sea is entirely free from pollution.

The Black Sea represents a particularly acute problem because of its enclosed basin and its stagnant hydrological stratigraphy. The river-borne pollution entering the sea increased tenfold between 1950 and 2000. Plants that produce oxygen by photosynthesis have been killed on the margins, and eutrophication has resulted in thick tangled masses of algae that kill fish and wash up on beaches. The deeper levels of the sea are anoxic and poisonous. In addition, an exotic species, the Atlantic comb jellyfish, which feeds on fish eggs and larvae, was introduced by a ship emptying its ballast tanks in the early 1980s and has experienced a population irruption. The fish stocks of the Black Sea have declined by 90 percent as a result.

AGRICULTURE

Mass-production industrial agriculture became an important presence in the northern Mediterranean after World War II. Millions of intensively reared cows and pigs in feedlots produce nitrogenous waste that in many cases flows into the rivers and the sea. The Po flood plain in northern Italy has a concentration of these factories, with eleven million animals in 1990 producing an amount of organic pollution three or four times greater than the sewage from the sixteen million humans who inhabit the same region. A large amount of arable land is devoted to raising food for these artificially fed animals. Processing of vegetable material can also produce serious pollution. The Saronic Gulf near Athens receives fifty thousand tons of waste from olive oil production alone. A stream flowing by the Temple of Hera on the island of Samos, where mythology says the goddess bathed frequently to renew her virginity, is so filled with olive waste that it appears to be a stream of rancid olive oil, not water. Hera would not be pleased to bathe there now.

Salinization remains a problem; along with waterlogging, it occurs in the Guadalquivir Valley in southern Spain and along the Ebro River in the northeast of the same country. It is a process of ancient importance in Iraq and in many parts of Egypt, including the Fayum depression, which being below sea level has no drainage and must depend on evaporation.

Industrial agriculture has greatly increased pollution by wasteful methods, such as spraying crops with pesticides at intervals to kill pests that may not show up at all, or setting irrigation schedules without regard to weather. A method that threatens a new cycle of erosion is the use of bulldozers to scoop out false terraces without retaining walls. When the returns from agriculture are low in an area where land development is taking place, the land is almost invariably sold and subdivided for industrial or touristic uses, so the area available for food production decreases. In the south and east, population pressures are forcing subsistence agriculture onto fragile marginal land where the soils can easily be exhausted. This is one of the worst cases of nonsustainability, since peasants who fail in one such area will usually move to another equally marginal place. The area subject to erosion has steadily increased due to loss of vegetative cover; in Greece, for example, surveys have observed erosion on 40,000 square kilometers (15,400 square miles), or about 30 percent of the country. The processes of industrialization and market integration have been particularly corrosive of the way of life of peasants who live in marginal areas such as mountains in the Mediterranean region.

Modern agriculture could be improved by emulating natural ecosystems, using biological controls, local manure, and drip irrigation where appropriate. Such methods based on scientific ecology have been used successfully in arid regions bordering the Mediterranean, such as the Negev Desert in Israel. Israel recycles about 40 percent of its sewage for use in agriculture. Agricultural improvement is also closely related to the prevention of deforestation, discussed below. As Lindsay Proudfoot and Bernard Smith note, "The key to soil erosion has been the destruction of the mature forest cover in conjunction with the natural erosivity of Mediterranean environments produced by steep slopes and intense seasonal rainfall" (Proudfoot and Smith 1997, 301).

Pastoralism continues to be important in the Mediterranean region, especially in the south and east, although it makes up an ever-smaller proportion of the total economy. Sheep are an important source of production, although they have become less numerous in Spain, where the Mesta has lost much of its former power. Goats continue their voracious use of the vegetative cover. Greece, for example, had 5.9 billion goats in 1996, which amounted to 50 percent of the entire goat population of the European Union. An interesting detail is the decline of the water buffalo population. They are still commonly seen in Iraq and the Nile Delta in Egypt, but rarely on the north side of the Mediterranean; their number in Greece dropped from more than 100,000 in the 1950s to 900 in 2000, with elimination of even that small number seemingly imminent with the loss of wet meadows due to the encroachment of cultivated fields.

WATER MANAGEMENT

Water is the "Mediterranean basin's most critical resource for the future" (Thornes 2001, 261). The need for water for burgeoning urban centers, but even more for food production, now approaches the amount of freshwater available in rivers and wells. For example, the water diverted annually for agricultural purposes from the Po River in 1990 was eighteen billion cubic meters, amounting to 36 percent of Italy's entire water use, and the amount has since increased. In the summer, the river's bed becomes completely dry in places. Water demand in North Africa and the Near East is already at least 90 percent of the available supply.

Major projects have been designed for water transfer for irrigation, such as the Southern Conveyor canal in Cyprus, which brings water from the Troödos

Mountains to the eastern lowlands. Another is the Libyan project to bring about 9 million cubic meters per day of fossil water from aquifers under the Sahara to coastal regions in the north, which means depleting a nonrenewable resource. Libya's water picture is bleak because there are no permanent streams or rivers. The Egyptian Aswan irrigation scheme is discussed in the case study in Chapter 7. Water problems become highly politicized when watersheds are divided among competing entities, such as the Jordan River, which drains a watershed divided among Israel, Palestine, Jordan, and Syria; the Euphrates-Tigris system, which flows through Turkey, Syria, Iraq, and Iran; and several rivers that flow from Spain into Portugal. Greece's four large northern rivers, the Axios, Strymon, Nestos, and Evros, all have the major proportion of their upstream watersheds in neighboring countries: Bulgaria, Turkey, and the former Yugoslav Republic of Macedonia. In Egypt's case, the upstream states are outside the immediate Mediterranean zone (although of course within the Mediterranean watershed), but the issues involved are every bit as critical. As demand rises and supply remains the same, or declines due to the weather pattern shifts associated with global warming, it is quite conceivable that the Mediterranean, African, and Near Eastern area could witness wars over water.

DESERTIFICATION

The crisis in water supply is closely related to the increasingly recognized problem of desertification in the Mediterranean basin. The United Nations Environment Programme defines desertification as "land degradation in arid, semi-arid and dry subhumid areas resulting from various factors including climatic variations and human activities" (Geeson, Brandt, and Thornes 2002, 5). Beginning in 1991, the European Community established and funded a program called Mediterranean Desertification and Land Use (MEDALUS), with the cooperation of many universities in the area, to gather information and to study certain areas in the Mediterranean basin that were judged to be examples of desertification or places in danger of desertification. The International Convention on Combating Desertification, ratified by more than fifty nations since 1996, recognizes the distinct desertification problems of the Mediterranean countries, all of which contain areas of severe water deficit. Studies in the late 1990s revealed that the Mediterranean area suffered a decrease in annual average runoff of 30 percent in the two previous decades. During that period, the growing demands for water caused by tourism and agriculture were facing a water resource shortage made worse by meteorological drought. The dilemma does not seem to have ended. Hazel Faulkner and Alan Hill are undoubtedly correct in stating that "the evidence suggests that the region is becoming increasingly 'desertified'" (1997, 270).

Two peculiarities of the Mediterranean climate exacerbate the problem: the lack of rainfall between May and October, and the extreme variability of rainfall from one year to another. Both of these anomalies represent insecurity for vegetation generally and agriculture in particular, and are pressures leading to widespread dependence on irrigation. In Greece, 80 percent of water is used for irrigation, in Italy 50 percent, in Spain 68 percent, and in Portugal 52 percent. Dry farming can be used only with crops such as trees that are drought-resistant, or with winter crops such as winter wheat. Paradoxically, flooding is also a problem in the same areas because rainfall, when it comes, is often torrential, and practices of soil compaction and deforestation increase the runoff during storms. The result is often water wastage as the floods sweep down into the sea before they can be used. The MEDALUS report concluded "that there is a strong interaction between water supply and erosion-vegetation cover and that to appreciate and solve the water resources problem the more conventional desertification problem of land degradation has also to be solved" (Brandt and Thornes 1996, 545).

Desertification presents different aspects in the two halves of the Mediterranean basin. In the south, it has often been seen as the northward advance of the Sahara on a long front, threatening to one degree or another all the nations on the north coast of Africa. Districts on the southern margin of the Atlas Mountains that were formerly forested or rich agricultural landscapes are now bare and desiccated. Tunisia, however, has found a way to reverse the process at least in one place. Oil exploration had inadvertently discovered a water source with tremendous pressure in 1972 south of the depression of Chott el Jerid, and this was used to create the artificial oasis of Rjim Maatoug with a town and important agricultural development watered by wells and protected by barriers against the blowing desert sands.

In the north, the spread of desertification is more like that of a disease that manifests itself in patches that appear depending on local conditions, and then may spread and join. United Nations studies have identified dryland soil degradation in parts of Spain, Sardinia, Sicily, Greece, and Turkey (United Nations Economic Programme 1992, map 11). The drier parts of Spain have been undergoing a process of desertification for many decades. This has been caused by heavy grazing, as noted previously, woodcutting, and wind and water erosion. Destruction of the native vegetation and erosion of thin soils have had a devastating effect on plant cover and on the productivity of the land.

LOSS OF BIODIVERSITY

Biodiversity as a word and as a matter of discussion among ecologists and environmentalists first appeared in the mid-1980s. It refers to the number of species, the complexity of ecosystems, and the variety of habitats in the environment. Some degree of variety and heterogeneity is necessary for the existence of most forms of life, including human life. Many ecologists believe that biodiversity provides buffers or functional resilience to the ecosystem, that is, that an ecosystem with a variety of species can better resist some forms of damaging change than an ecosystem with only a few species. An attempt to preserve species, therefore, should be intended not simply to save the intrinsic value of that one species, but more importantly to save its function as an interactive constituent of the entire environmental structure.

The case could be made that biodiversity came under assault in the Mediterranean and Near East earlier, and therefore suffered more damage sooner than in any other part of the world. The only other area that is somewhat comparable in that respect, again, is China. There is a long list of species that once existed in the Mediterranean that are now missing or extremely rare, from lions to the medicinally valuable silphium plant. There are no longer any wild crocodiles, hippopotami, or sacred ibis in Egypt. The brown bear, a relative of the American grizzly, now survives in the Mediterranean area only in isolated mountain retreats and in national parks. The last Iberian lynx in Portugal was shot in 1892, the last one in the Pyrenees in 1902, and only a few survive in Las Marismas in southern Spain. Barbary apes are limited to a few colonies on the coast of North Africa and in Europe only on the rock of Gibraltar. In Greece, the Cretan Bezoar goat, or agrimi (*Capra aegagrus cretica*), has only a small remnant population in the White Mountains of Samaria on Crete. Although nominally protected inside Greece's most popular national park, it is nonetheless endangered due to poaching and crossbreeding with feral domestic goats.

An unfortunate fact is that in the contemporary period, loss of biodiversity is still occurring, indeed more rapidly than ever. Factors that cause this loss include the destruction or transformation of habitats including but not limited to forests, hunting or fishing in an unsustainable manner, pollution, and the introduction of exotic species that attack or compete with native species. Sometimes the destruction is so complete that even if an area could be protected, nothing like the original ecosystem would regenerate. Endemic species, which exist in one small area and not in any others, are in great danger of extinction. To give one example, Eleonora's falcon, endemic to the Mediterranean climatic zone, survives only on rocky isolated cliffs. The island of Crete has 130 species of plants that exist only there, about one-tenth of all of its native plant species.

Human interference has caused reduction of the numbers and ranges of native plants and animals, and the extinctions of many. For example, in Spain 37 percent of the bird species are either endangered or vulnerable, along with 53 percent of wild mammals, and in Italy, half of all reptile and amphibian species are so. Predatory birds such as eagles, hawks, and vultures have exhibited exceptional decline in numbers. These birds, along with seabirds, have suffered disruption of their reproduction due to insecticides. But small songbirds also suffer chronic losses from the fact that scores of millions are shot and captured each year as a delicacy for the table or for sale in markets, a culinary preference characteristic of some Mediterranean peoples. In Malta, an important stopping place for migratory birds, three million are killed annually. "Some of the best Mediterranean pâté is made from the flesh of the blackbird and the thrush, and thrushes, particularly, turned on a spit, are a favourite winter lunch in many parts of the south of France," Richard Carrington observed (1971, 143–144). Attempts to deal with other environmental problems may inadvertently impact desirable species. In Israel, pumping for irrigation has virtually destroyed the habitat of the otter, which has been reduced to a relict population of a few dozen.

Among introduced species are domestic animals, many of which have the ability to become feral and spread into wild areas. In parts of the Mediterranean, ecological disruption has gone on for so long that it is often difficult to tell which animal or plant species are natural and which are introduced. Some are clearly foreign, such as the American prickly pear cactus, which has spread widely because it is resistant to grazing (but goats are known to eat it too). A danger to many forms of sea life in the Mediterranean is *Caulerpa taxifolia*, a plant used as an ornamental in fish tanks that was accidentally released into the sea from aquariums. In fact it was first noticed in 1984 as a small clump of plants on the seafloor near the Monaco Oceanographic Museum, and it has been determined that it is a mutant strain found in aquariums but not in the wild until it escaped. Within a decade it had spread widely; it now blankets rocks and sea bottom along hundreds of miles of the Mediterranean coast and continues its rapid proliferation. It contains a poisonous sap that ensures that no fish will eat it; indeed, thus far only one organism (a sea slug) has been found that will eat it. The area it covers has been described as a biological desert.

Whales and dolphins are in particular danger in the Mediterranean, and their numbers are believed to be declining every year. There is no legal hunting of them, although the nations that have laws to protect them have not taken enough efforts to ensure they survive. They are killed by various pollutants and diseases; thousands of striped dolphins died of a disease in the western Mediterranean in the early 1990s, and many of them washed ashore. Like land animals, they perish when their habitat is impaired. Overfishing reduces their food source, causing starvation, and fishermen often kill them in cruel ways by shooting them or severing their tail flukes because they consider them competitors for a dwindling supply of fish. Ironically, they are friendly creatures credited in mythology with the rescue of shipwrecked sailors.

The Mediterranean monk seal is one of the most endangered species in the world. It received its name from the fact that it seeks refuge in caves, often with underwater entrances, and on rocky coasts isolated by cliffs. But it was in early times a gregarious species that congregated on the same beaches that tourists now enjoy, and it escaped to the retreats that the remnant population occupies because of human persecution. Small communities of monk seals survive on the coasts of Greece, Turkey, Morocco, and Cyprus, and on the smaller islands near Madeira, but the total surviving number is believed to be limited to about three hundred. In order to save this gentle creature, refuges must be established and fishermen convinced not to kill them just because they eat fish and are therefore competitors, although they are now so few that they cannot seriously be considered depleters of fish populations.

Sea turtles are also endangered species in the Mediterranean. Nesting females probably number under 2,400, down from the hundreds of thousands that existed at the beginning of the twentieth century. Many of the female loggerhead turtles nest on beaches on the Greek island of Zakynthos and in Turkey near Dalyan. A turtle refuge has been established for the Greek population, but there has been strong opposition by some people on Zakynthos because the refuge interferes with the profitable development of the island for tourism. Cyprus protects Akamas Beach, a turtle nesting ground. In the formerly undisturbed eastern Mediterranean, tourists seeking clean, uncrowded beaches away from the degraded ones in the west are interfering with the eggs and hatchlings. At the same time, tourists generally produce more waste and litter than residents, and tend to deposit it in more hazardous places such as beaches. Sea turtles mistake plastic bags for jellyfish and die a slow death when the plastic obstructs their digestive systems. In addition, tens of thousands of turtles, including many that are visitors to the Mediterranean from Atlantic populations, are caught every year on the hooks of long-line fishermen. A turtle rescue program at the Naples Aquarium is able to save about twenty-five animals per year.

FISHING

Although the Mediterranean is not as rich in fish and other marine life as the world's larger oceans, it does possess many species, some of which are unique to

Fishing is a major industry affecting the Mediterranean maritime environment. This fishing boat in the harbor at Brindisi, Italy, illustrates the equipment of a small-scale fishing enterprise. (Photo courtesy of J. Donald Hughes)

its waters, in numbers sufficient to support a fishing industry in most of the surrounding nations, and demand for seafood and other products such as coral and sponges means that the sea's living resources are persistently exploited. Mediterranean fish are highly prized and fetch prices to match; they make up 5–6 percent of the world market. Unfortunately, diseases caused by pollution have been increasingly reported. A significant number of species of fish are in danger of extinction within the Mediterranean.

Before World War II, most fishermen around the Mediterranean operated small sail and row boats, fishing with long lines, traps, drifting gill-nets, entangling nets, bottom-set nets, and small seines. Trawlers using purse seines to capture shoals of fish, which they attract by lights, became more common in the 1950s, but they were small, perhaps averaging fifteen meters in length and powered by engines of thirty to sixty horsepower. As high technology became available in the late twentieth century, many were equipped with echo-sounders, global positioners, and other sophisticated electronic equipment to aid navigation and locate fish. Boats with much more powerful engines towing pelagic trawls, fishing gear that enables them to fish in midwater and capture anything that swims—including adult and juvenile fish and even dolphins—are now widely operated. A great variety of other methods are used, including drift nets seven to fourteen kilometers (4.3 to 6.2 miles) long and up to thirty meters (100 feet) deep, which kill not only the target species but also other fish, turtles, whales, dolphins, and seabirds. Drift nets are now banned by the European Union because of their decimation of fish stocks, but Italy, Turkey, and France continue to operate noncompliant ships. The most damaging drift net fleet is the Moroccan, which operated 177 boats in the Alboran Sea in 1988, the westernmost basin of the Mediterranean, where as a side-effect of its search for swordfish it kills from three thousand to four thousand endangered short-beaked common dolphins annually (they are not all that common; this number is more than 10 percent of the total population), along with sea turtles and other rare species. Only some species of fish, of a certain size, will sell at prices that enable the trawlers to pay their operating costs, so a large part of the catch is dumped back, dead, into the sea, where it attracts screaming crowds of gulls and other seabirds.

Several nations have adopted legislation limiting the number of trawlers and the fishing seasons in an attempt to conserve the resource. Some also regulate sport fishing. The enforcement of these laws varies around the Mediterranean from strict to lax. As a result, for example, trawlers from Sicily, where aggressive fishing has depleted the stocks, invade Tunisian waters where conservation has been more effective and are sometimes arrested. In addition, foreign fishing fleets from Asia, especially Japan, and elsewhere enter the Mediterranean to exploit its international waters, where the states have not established exclusive economic zones extending beyond their territorial waters like those that exist in the oceans. The Mediterranean nations have little or no control over what these foreign fleets do, and the interlopers avoid regulation in any case by often changing their names and flags of convenience. The numbers of fish that can be caught, as well as their average size, have declined sharply. Fishermen off Sardinia used to enclose shoals of tuna with their boats and nets, but now there are no more big catches. Meanwhile, the health of the surviving fish deteriorates from pollution and diseases. Without effective international cooperation, the collapse of Mediterranean fisheries is a genuine possibility in the near future.

FORESTRY AND DEFORESTATION

There is a contrast between the European northern Mediterranean nations on the one hand and those in Africa and Asia (the south, including the eastern

A forest fire rages on a hillside below a church in the village of Vavla, Cyprus, June 16, 2000. Devastating forest fires inflicted one of the Mediterranean island's worst ecological disasters in decades. (Reuters/Corbis)

Mediterranean with the exception of Israel) on the other as regards forestry and deforestation. In the north, forests increased over the recent period by about 0.7 percent annually, although this includes some marginal growth of maquis. Greece had a forest cover of 19 percent in 1987. A new form of deforestation, clearing mountain slopes to create ski areas, invaded the Alps, Pyrenees, Apennines, and some of the Greek mountains as alpine sports grew in popularity. Also, "forest death" due to acid precipitation is thinning many of the existing forests, as noted below.

In the south, however, deforestation continues, and in the fourteen Mediterranean Afro-Asian countries the forest cover now amounts to only 6.6 percent. In Israel, a replanting program has increased the area of forest cover in the northern part of the country to 10 percent. Over the region, however, the area of forest lost between 1920 and 1980 amounted to thirteen million hectares. In many local areas, forest stands are in danger of complete removal due to urbanization, industrialization, and conversion to other uses. Virtually all of this forest decline is due to human actions, and it is because of it that so many observers have noted that a large proportion of the southern Mediterranean and the Near East appears to be exhausted land. It is difficult to find a healthy mature closed forest stand anywhere in the area. All that exist are in danger.

Periodic fires are natural in most Mediterranean environments, and plants have evolved physiological strategies to enable their survival after fire. Less intense fires may spare the largest trees and stimulate regrowth. But human interference has altered the fire regime. Usually this consists of deliberate setting of fires, shortening the natural cycle and transforming the forest into a more degraded form of vegetation. In the second half of the twentieth century and the early twenty-first century, the use of arson in the forests became a form of political protest. Studies of the frequency of fires in the forests of Greece have shown that they increase in number and severity in the weeks before elections. In Israel, the problem of deliberately set fires increased many times after the beginning of the *intifada*, a protest movement of Palestinian Arabs against the occupation of the land by Israelis. Sometimes, however, protection of forests means the total exclusion of fire and lengthening the natural burning cycle so that fuel accumulates to the point where fire, if it occurs, may be so intense that all vegetation is destroyed and the soil rendered almost sterile. Since humans are the major force altering the forest, it would seem that efforts should be made to emulate a moderate natural fire regime that would maintain the forest types native to the region. Financial and labor resources to accomplish this are lacking, however, in all but a few Mediterranean nations.

TECHNOLOGY

Although the new technology of the internal combustion engine and petroleum transformed Europe from the early decades of the twentieth century, there were many parts of the Mediterranean that did not experience its full impact until after World War II. Hundreds of thousands of Greeks first learned to drive automobiles (*aftokiniti*) in the 1960s, and in cities further east and south traffic jams caused by cars came even later as everyday phenomena. They came with a vengeance, however, made worse by the antiquated street plans when they arrived. In 1978, the number of private cars per one thousand people was 11 in Turkey, 80 in Greece, 178 in Spain, and 300 in Italy. Those numbers were, of course, increasing rapidly in each of those countries.

One of the most destructive changes in technology was the invention and widespread availability of small, easily handled, and very powerful tools. The gasoline-powered chain saw was manufactured by Andreas Stihl in Germany in 1929, but lightweight models that one person could wield did not gain wide distribution until after World War II. Before the chain saw, two loggers with a hand-drawn crosscut saw might have taken two hours to fell a tree that one man with a chain saw can now take down in two minutes. The effect of this tool, now available everywhere and affordable by all except the poorest, on the forests of the Mediterranean has been catastrophic.

Another invention that has been horrifyingly destructive is the production of a series of portable, handheld guns capable of firing a large number of bullets in a very short time. These automatic weapons, designed unfortunately for killing human beings, are also used by poachers to kill large wild animals. A herd can be mowed down all at once; the last wild group of Arabian oryx was slaughtered in this way, although later an effort for reestablishment of the species in Saudi Arabia succeeded with the use of animals from zoos in the West. Soldiers armed during wars have been known to shoot any wildlife in sight as a degenerate form of recreation. Fishermen at sea use them to kill sea mammals to prevent their competition for fish.

Extractive technology uses huge machines that greatly increased the speed of earth removal and consequently the extent of damage to the landscape by strip mining. Even more intimidating has been the increase of scale in the technology of war, including more destructive explosive devices that could be undiscriminating in their effects on nature as well as the works of humankind—and humankind itself. Nations tend to regard war as necessary in their interests, and they rarely exercise restraint for environmental reasons.

CONSERVATION

Working to save the Mediterranean Sea and the adjoining lands from further degradation, and attempting to restore a livable environment, has been the goal of many governmental and nongovernmental organizations in recent years. The environmental movement was perhaps slow to get started in Mediterranean countries as compared with northern Europe and the United States, but by the 1970s there were local groups at work on conservation projects and on attempts to get environmental laws passed, as well as local chapters of international environmental groups. For example, in Greece the Hellenic Society for the Preservation of Nature was incorporated in 1951. The Society for the Protection of Nature in Israel was also founded in the 1950s. In Italy, an organization called Italia Nostra ("Our Italy") was founded in 1955 and centered its efforts on preserving the Italian cultural heritage, including within that purview many aspects of the natural heritage of the country. An Italian section of the World Wildlife Federation was organized in 1966. Almost every Mediterranean country now has a number of such associations of citizens. They are particularly active in France, Spain, Italy (where conferences have been held in Assisi, the town associated with St. Francis, the medieval forerunner of environmentalism), Greece, Turkey, and Egypt, and have more recently appeared in Algeria and Tunisia. Some of these movements became involved in antipollution and antinuclear efforts in the political sphere. The environmental movement in France, where nuclear power generation reached its most extensive and intensive development, understandably gave major attention to antinuclear concerns. "Green" political parties have been formed in some of the democratic nations, although they have not been as successful as in northern Europe, notably in Germany.

A major achievement of international environmental diplomacy on the governmental level was the Mediterranean Action Plan (MAP), negotiated in 1975 by all the then-existing states of the Mediterranean littoral except Albania (which later acceded to the contract), and formalized by the 1976 Barcelona Convention for the Protection of the Mediterranean Sea Against Pollution. The Blue Plan, as it was called at its inception, was intended to

> make available to the authorities and planners of the various countries in the Mediterranean information which will enable them to formulate their own plans to ensure optimal socioeconomic development without causing environmental degradation . . . and to help the governments of the states bordering the Mediterranean region to deepen their knowledge of the common problems facing them, both in the Mediterranean Sea and its coastal regions. (Grenon and Batisse 1989, vii)

"The environment is our home," announces this slogan on a bus in downtown Athens. Environmental organizations in nations around the Mediterranean Sea seek to arouse public concern by messages such as this one. (Photo courtesy of J. Donald Hughes)

The United Nations Environment Programme (UNEP) facilitated the proceedings and continues to operate the secretariat, the Coordinating Unit for the Mediterranean Action Plan, in Athens. The objectives of the agreement include facilitating cooperation among the Mediterranean states in order to combat massive oil pollution, assisting information exchange and technological cooperation among nations, and helping states develop their own national programs. MAP envisioned extending its efforts to control the dumping at sea of other pollutants in addition to oil.

A further agreement under MAP in 1980 set standards for the control of pollution of the sea from land-based sources. It listed the substances to be controlled, established a process for setting precise levels of permissible concentration, and promised technical assistance to help the developing countries of the basin to implement their efforts under the agreement. A Mediterranean trust fund was created, supported in large part by the European Economic Community (now the European Union). The program supports scientific research and development planning and has assisted in the construction of sewage treatment plants in a number of cities, including Marseille, Athens, Thessaloniki, Aleppo, Alexandria, and Cairo. Regrettably, standards by 1990 had been set for only eight out of fifty measures of pollution control, and enforcement of those agreed upon has been spotty. Although MAP's budget is very small, some of the signatory nations have been slow to pay their assessments for its support.

This international effort nonetheless is an amazing achievement considering the often-antagonistic dispositions of the nations around the inland sea, and the fact that it was negotiated during the period of the Cold War when the political tides of the Mediterranean felt the pull of the competing interests of the United States and the Soviet Union. To the extent that it succeeded, it can serve as a possible model for other regional agreements among the Mediterranean states. After all, this is an organization that has enabled cooperation between such unlikely combinations of nations as Greece and Turkey, and Israel and several Arab states, all of which have occupied seats at the same table to tackle this environmental crisis.

RELIGION AND THE ENVIRONMENT

In recent times, each of the great monotheistic religions that together claim the allegiance of the great majority of the Mediterranean peoples, namely Islam, Roman Catholic and Eastern Orthodox Christianity, and Judaism, has tried to come to grips with the ecological crises that face humankind, and to apply the resources of scripture and tradition to them. All three affirm the goodness of the natural world as the creation of God, and the duty of humankind to care for it as God's steward on Earth. But each of them has a characteristic emphasis of its own.

Islamic scholars have given increasing attention to ecological problems during the previous generation or so. One of the most noted of these students of religion and science is Professor Seyyed Hossein Nasr, who explains that the central axis of the Islamic worldview is unity (*al-tawhîd*). This sense of unity that pervades all things in nature derives directly from the unity of God. As the Persian poet Sa'di expressed it, "I am in love with the whole universe because it comes from Him." Islam, says Nasr, created an advanced science between the ninth and sixteenth centuries, without which Western science could not have developed, but Islam never had a secular science independent of a spiritual vision of the universe. Islamic ecology, therefore, is a study of the whole environment as a complex unity in which everything is interrelated, including the psychological and spiritual levels. In the Islamic view, man is the vicegerent of God on Earth (*khalîfat Allâh fi'l-ard*). The profanation of nature through its so-called conquest occurs, therefore, when man attempts a futile separation of the terrestrial world from transcendent reality, which constitutes a "revolt against heaven." If man is true to his spiritual character, he acts as a channel of grace for nature.

But how does this theological insight, based on the Qur'an, work out in practical dealings with the natural environment? Starting from the same affirmations as Nasr, Iqtidar H. Zaidi points out that what is needed is man's stewardship in the management of resource processes, as a temporary guardian, so as to enrich rather than degrade the quality of the environment:

> *That ye exceed not the measure,*
> *But observe the measure strictly, nor fall short thereof.*
> *And the earth hath He appointed for [His] creatures. (Qur'an 55:8–10)*

Thus God forbids environmental disruption. Human greed and mismanagement must be controlled. Zaidi maintains that the way this should be done in practice is that the Islamic state must control people's actions so as not to allow the environment to deteriorate. The authorities of the Islamic state are instruments of God's wrath. Justice does not permit the exploitation of resources by one human being in a way that harms another, including the impairment of environmental quality. The model of the proper treatment of the environment in an Islamic society, therefore, is a state that enacts positive environmental laws in accordance with Qur'anic principles and enforces them in an equitable manner.

The First Islamic Conference of Environment Ministers met in Jeddah, Saudi Arabia, on June 10–12, 2002, to conclude a declaration for presentation at the Earth Summit (the United Nations World Summit on Sustainable Development) to be held later that summer (August 26 to September 1) in Johannesburg, South Africa. The meeting of environment ministers was the culmination of several other Islamic forums on the environment. The declaration began by reaffirming that man is the lieutenant of Allah on Earth, responsible for harnessing and protecting the environment. (Despite the declaration's use of "man" in sentences like this one, it clearly states that women shall be recognized as full partners in sustainable development.) The care of the environment in which people live is an expression of the primary Muslim virtue of charity. Since the environment is a gift from Allah to humankind, individuals and communities have a duty to care for it and refrain from acts that would result in pollution or damage to the ecosystem, or disturb its balance. Among the factors named as interfering with nondamaging sustainable development are poverty, public debts, desertification, wars, overpopulation, overutilization of natural resources, absence of nonpolluting technology, and insufficiency of expertise.

Among the practical measures urged to meet these twenty-first–century challenges were securing investment and funding sources, improving educational and health programs, the transfer of environment-friendly technology and research, preservation of the civilizational heritage, and securing effective participation of developing countries in decision-making.

Christian leaders of the Roman Catholic and Greek Orthodox communities agree on some points and differ on others, but in recent times they have identified a broad area of consensus on humanity and the natural environment. Following a symposium of some two hundred Catholic and Orthodox environmental scholars that was held during a cruise on the Mediterranean Sea among Corfu, Greece, and Venice, Italy, Pope John Paul II and Ecumenical Patriarch Bartholomew of Constantinople signed a "Joint Declaration on Articulating a Code of Environmental Ethics" on June 10, 2002, simultaneously during a live television link. "It is on the basis of our recognition that the world is created by God that we can discern an objective moral order within which to articulate a code of environmental ethics," states the text of the historic agreement, which goes on to declare that Christians are "called to collaborate with God in watching over creation in holiness and wisdom." The declaration voiced concern about the negative consequences for humanity and for all creation resulting from the degradation of basic natural resources such as water, air, and land, "brought about by an economic and technological progress which does not recognize and take into account its limits." Almighty God, the authors continued, envisioned a world that would embody beauty and harmony, and created it, making every part of it an expression of freedom, wisdom, and love. God placed human beings at the center of the whole of creation. Humans share many features with other living beings, but God also gave them an immortal soul, with self-awareness and freedom, endowments that give them the image and likeness of God, including the ability to cooperate with God in realizing ever more fully the creation's divine purpose.

Unfortunately, humans disobeyed God at the beginning of history, rejecting his design for creation and destroying its harmony. "If we examine carefully the social and environmental crisis which the world community is facing," the declaration continues, "we must conclude that we are still betraying the mandate God has given us: to be stewards called to collaborate with God in watching over creation."

In order to overcome this tendency and to obey the mandate, the pope and patriarch affirmed certain ethical goals, which follow: We should consider children and the coming generations whenever we evaluate our options for action. We should base our values on natural law. We should use science to correct the mistakes of the past and enhance the well-being of present and future generations, but not just materially. Human and spiritual values and the common good must be central. We are not the owners of the Earth; we are only stewards of the common heritage, and in that light must be careful about our actions.

In taking responsibility to work for a better world environment, not every person and institution must assume the same burden:

> Everyone has a part to play, but for the demands of justice and charity to be respected the most affluent societies must carry the greater burden, and from them is demanded a sacrifice greater than can be offered by the poor. Religions, governments and institutions are faced by many different situations; but . . . all of them can take on some tasks, some part of the shared effort.

The agreement recognizes that there may be disagreements about environmental issues, and it calls for mutual respect and open exchange in resolving them, concluding, "It is not too late. God's world has incredible healing powers. Within a single generation, we could steer the earth toward our children's future. Let that generation start now, with God's help and blessing."

Recent statements by Jewish scholars have generally reaffirmed the traditional framework outlined in Chapter 3: the biblical statements about the Creation and the commandments protecting the Earth and its creatures, and the tendency of Jewish tradition to extend such principles as "do not destroy" (*bal tashhit*) beyond the enemy's trees mentioned in the Torah (Deuteronomy 20:19) to any natural object of possible human use. The grant of "dominion" to humankind is generally taken to indicate not license to trample, but a charter of responsibility. In Judaism, as in Islam and Christianity, human beings stand as the stewards of God, placed on Earth as Adam was in Eden, "to till and to look after it" (Genesis 2:15). The question can be asked in regard to any of the three religions, "How well have they done this?" The answer in each case must be a qualified one.

In considering Judaism and care of the environment in the Mediterranean area, the instance of Israel inevitably comes to mind. Undoubtedly many of those who founded and managed the new state in the latter half of the twentieth century had biblical precedents in mind. The land that came under Israeli administration was, it must be admitted, selected as the modern homeland because it was the biblical locale, the "land of milk and honey" (Exodus 3:8). Israelis established a deep tie to the land, and such environmental programs as tree planting were enthusiastically embraced from the beginning. Wildlife refuges were established. But it is only in a very limited sense that Israel can be regarded as a place where an attempt has been made to carry out the environmental prescriptions of the Bible and Jewish tradition. There is no doubt that development has been emphasized at the expense of environment, and that

Tourists at the Ein Gedi Nature Park, by the Dead Sea, Israel. Visiting protected areas like this one for the purpose of ecotourism is now a widely recognized form of recreation. (Richard T. Nowitz/Corbis)

many serious environmental errors have been made. Taking water for vast new agricultural projects has depleted and polluted many streams and reservoirs. Chemicals have been spread on the land by farms and factories. Roads and other infrastructure in many places have needlessly marred the landscape, and wars and violence have caused scars. But Israel has also developed an active environmental movement, and contemporary environmental concerns are voiced there from both religious and secular points of view.

MODERN IRAQ: ENVIRONMENTAL PROBLEMS

The history of twentieth- and twenty-first-century Iraq has been marked by a series of environmental problems that are pressing and severe. Some of these were originated or exacerbated by the effects of military conflicts, including World War I, the Iran-Iraq War, internal battles involving the Kurds and Shiite Muslims, and the Gulf Wars of 1991 and 2003. Others may be said to result from negative policies of Iraqi governments; the impacts of economic sanctions imposed by the United Nations, the United States, and its coalition partners; and failure to achieve cooperation with regional neighbors, especially Syria, Turkey, and Iran, on the management of shared natural resources such as water and oil.

To begin with the political history, the Ottoman Empire dominated Iraq at the start of the twentieth century, but lost that province as the result of World War I, during which a British campaign in Mesopotamia beginning in 1914 captured Baghdad in March 1917 and Mosul a year later. At the end of the war Britain received Iraq as a mandate from the League of Nations, and in 1921 it imposed a constitutional monarchy under Amir Faisal. Independence was granted formally to the Arab kingdom in 1932. The monarchy continued until 1958, when a military coup installed a republic. The Ba'athist Party took power five years later, and Saddam Hussein became leader of both party and nation in 1979. Saddam used conventional and chemical weapons against Kurdish insurgents, including civilians, in the northern region of Iraq. Iraq fought an extremely bloody but ultimately unsuccessful war against Iran in 1980–1988, with United States support for the Iraqis that gradually faded. In the war against Iran, Iraq used chemical weapons including mustard gas and the nerve gases sarin, tabun, and GF, with losses of human life in the tens of thousands and accompanying environmental damage. The use of biological weapons remains unconfirmed. Iraq invaded the oil-rich nation of Kuwait in 1990, provoking the Gulf War of 1991 and the military action sanctioned by the UN, including the incursion of forces from the United States and more than thirty other nations into

Kuwait and southern Iraq, which ended with a cease-fire on March 3, 1991. Much environmental damage resulted from the systematic destruction of Iraq's infrastructure and the vindictive actions of the Iraqi military, including setting fire to more than six hundred Kuwaiti oil wells and deliberately causing oil spills. By Security Council Resolution 687, the UN established UNSCOM, a special commission responsible for inspecting and supervising the destruction of Iraq's weapons of mass destruction (biological, chemical, and nuclear weapons). Under cover of the cease-fire, Iraqi governmental forces suppressed rebellions of the Kurds in the north and Shiite Muslims in the south of the nation. The U.S.-led coalition ordered Iraq to end all military activity north of latitude 36 degrees north, including overflights, in order to provide a safe haven for the Kurds. This was extended to 33 degrees north, just south of Baghdad, in 1996 to counter an Iraqi military offensive against the Kurds. A second "no-fly" zone had been imposed in 1992 south of latitude 32 degrees north to protect the Shiites in southern Iraq. After impeding the work of UNSCOM in various ways, Iraq ended all forms of cooperation with it in October 1998. The UN staff was evacuated, and the United States and United Kingdom began a bombing campaign against Iraq's air defenses in December. Faced with a call from U.S. president George W. Bush for action against Iraq, Saddam Hussein allowed UN inspectors back into Iraq in November 2002. In March 2003, the chief UN inspector, Hans Blix, said that Iraq had accelerated its cooperation and asked for more time to verify compliance. The UN Security Council failed to agree on possible next steps to ensure Iraqi compliance. On March 17, President Bush gave Hussein forty-eight hours to leave Iraq or face war, and the second Gulf War began on March 20. U.S.-led coalition forces entered central Baghdad on April 9, and widespread looting broke out, which the U.S.-led coalition did little to control. Combat operations ended on April 14, but insurgent operations have continued against the U.S.-led coalition and Iraqi citizens cooperating with it. The environmental impacts of the war will be discussed further on, along with more general environmental situations.

One of the greatest environmental problems of Iraq is a chronic water deficit due to low rainfall and a high rate of evaporation that result in making its agriculture dependent on irrigation from the Tigris and Euphrates rivers. Rainfall is highly seasonal, coming almost exclusively in the late autumn and the winter, and amounts to less than 100 millimeters (4 inches) over most of the country. Evaporation, in contrast, is often 3,000 millimeters (118 inches) in summer, presenting the continuing danger of salinization. In 1970, half the irrigated areas in central and southern Iraq were subject to this form of degradation. The dependence of the land and its people on the two rivers is as great in modern times as it was for the ancient Sumerians, Babylonians, and Assyrians.

The need to feed a growing population continues to press on the availability of water. In 1920 the population of Iraq was less than 3 million; in 2003 it was more than 24 million. The capital alone, Baghdad, grew from 3.8 to 5.6 million between 1987 and 2002.

Like Egypt, Iraq is a downstream state. Both the Tigris and Euphrates rivers arise from headwaters in Turkey, and the Euphrates also flows through Syria before reaching Iraq. The Euphrates is formed in Turkey by the confluence of the Murat and Karasu (Blackwater) rivers and flows southeastward 2,700 kilometers (1,700 miles) to its junction with the Tigris; of this distance, 1,400 kilometers (870 miles) is within Iraq. Precipitation in the upper basin in Turkey provides more than 95 percent of the natural water volume of the river. Its annual flow is approximately 28 billion cubic meters per year. The Tigris is not as long as the Euphrates, with a total length of 1,840 kilometers (1,143 miles), 1,360 kilometers (840 miles) of which is inside Iraq. Somewhat more than half of its annual flow of 70 billion cubic meters per year originates from tributaries within the borders of Iraq, and almost all the rest from Turkey, predominantly in Kurdish-inhabited territories in both nations. The volumes of both rivers are usually at their highest during springtime floods, and their salinity, while variable, is relatively high. Agriculture areas are subject to salinization today, however, to an even greater extent than in ancient times. The Euphrates and Tigris join one another near Qurna in southern Iraq, and their waters flow onward to the Persian Gulf as the Shatt al-Arab, which forms part of the boundary between Iraq and Iran.

Water rights on the two rivers present a difficult problem of international politics. Virtually all the water of the Euphrates originates in Turkey and flows through Syria before reaching Iraq. The Tigris also has important watershed in Turkey and Iran. Iran controls the largest eastern tributary, the Karun, which flows directly into the Shatt al-Arab. Claims on the water have been the subject of international negotiations that have had a turbulent course and are not yet settled. In 1911 the Ottoman government invited Sir William Willcocks, a British hydraulic engineer with a distinguished career in India and Egypt, to prepare recommendations for water control in Mesopotamia. As a result, barrages were constructed at Hindiyya on the Euphrates and at Kut el Imara on the Tigris to regulate flow for irrigation. After a hiatus during the world wars, Iraq's royal government created a water development board and began work on a series of dams on the Tigris tributaries in the north. Since Baghdad is located on the Tigris, the danger of spring floods damaging the city was severe, and in the 1950s works were constructed to divert Tigris waters into the Euphrates through a lake in the Tharthar Depression as needed. In the following decade, reservoirs for hydropower, irrigation, flood control, and storage were completed on the Lesser Zab and the Diyala, tributaries of the Tigris in the Kurdish Mountains.

Syria built the large Euphrates (Tabqa) Dam between 1968 and 1973. It is an earth-fill dam 60 meters (197 feet) high and 4.6 kilometers (2.85 miles) long, and its reservoir, Lake Assad, extends 80 kilometers (50 miles) upriver. It generates 8.9 billion kilowatt-hours of electricity, supplying power even to remote villages east of the Euphrates, and irrigates a quarter of a million hectares. During the filling of the lake in 1975, the level of the Euphrates below the dam fell severely, so that Iraq considered it a threat and in retaliation massed troops and threatened war. Saudi Arabia mediated an agreement that allocated Syria 40 percent of the water (under a 1990 renegotiation of the agreement, the amount is 42 percent). Syria subsequently constructed two further dams on the Euphrates.

Turkey represents a greater threat than Syria by far to Iraq's water. The Southeast Anatolia Project (GAP), envisioned beginning in the 1950s, would include twenty-two dams and nineteen power plants online in the early twenty-first century, mainly on the Euphrates but also on the Tigris, and would irrigate 1.7 million hectares. It would generate more than one-fifth of all Turkey's viable hydropower. When completed, it could reduce the flow of the Euphrates River to Syria by 40 percent, and to Iraq by 90 percent. It began with two large dams, Keban and Karakaya, in the 1980s. Turkey and Iraq had initiated a Joint Technical Committee on Regional Waters in 1980, which Syria joined in 1982, but Iraq dismissed the committee in 1990 after Turkey completed the giant Atatürk Dam, one of the largest earth-and-rock-fill dams in the world, 184 meters (604 feet) high and 1,820 meters (1.13 miles) long. Its reservoir covers 816 square kilometers (315 square miles), with a capacity greater than the Euphrates's annual flow, and its hydropower capacity is more than ten times that of Syria's Euphrates Dam. Subsequent progress on Turkey's project has been slower than planned and is complicated by the Kurdish issue. Syria has somewhat more geographical leverage than Iraq with Turkey on water, since in the case of the Orontes River, which flows into the Mediterranean, Syria is the upstream state and Turkey the downstream state. The difficulty of the entire water issue is underlined by the statement of Turkish president Suleyman Demirel that Turkey has a right to do anything it wants with its rivers. Turkey has, however, unilaterally agreed to provide a minimum flow of 15.8 cubic kilometers per year to Syria in the Euphrates. Iraq, being downstream from Syria, must take account of the fact that if Syria were to receive only this minimum volume, Iraq would possibly receive even less.

Today in all, thirty dams in Iraq store many times more water than the annual flow of its rivers. On the Euphrates, the amount stored is five times the an-

Atatürk Dam on the Euphrates River in Turkey, completed in 1990. One of the largest earth-and-rock-fill dams in the world, 184 meters (604 feet) high and 1,820 meters (1.13 miles) long. Its reservoir covers 816 square kilometers (315 square miles), with a capacity greater than the Euphrates's annual flow. (Ed Kashi/Corbis)

nual flow. Iraq stores twice the annual flow of the Tigris in the Tharthar diversion reservoir. Iraq has six dams on the Euphrates and three on the Tigris, plus five on the Tigris tributaries. Undoubtedly there will be more if and when Iraq achieves a stable government. Iran has eighteen dams planned, under construction, or constructed, almost all on the Karun and its tributaries, which lie within Iran and have little effect on the main flow of the Tigris.

The desiccation of the southern marshes of Iraq is an ecological disaster related to the water issue. In the mid-twentieth century, there were some twenty thousand square kilometers (7,700 square miles) of wetlands in southernmost Iraq that had existed since the earliest recorded times. Much of the earliest evidence for human culture in Mesopotamia comes from this district, and about half a million Marsh Arabs (Ma'dan) lived there, constructing buildings resembling those of five thousand years ago of bundles of reeds; fishing from skin boats waterproofed with asphalt; raising rice, millet, and dates; and herding cattle and water buffaloes. The marshes were also home to wildlife, including otters, boars, fishes, reptiles, amphibians, and millions of waterfowl such as cranes, storks, and ducks. In 1951 the British Haigh Report recommended draining the marshes to create land for agriculture, and soon thereafter a major drainage canal, the "Third River," began. Meanwhile, the spread of water hyacinth (*Salvinia*), an aquatic weed introduced from South America, clogged many of the waterways. Work languished until 1980, when it began again. In the first Gulf War, of 1991, the Marsh Arabs, who are Shi'ites, opposed Saddam Hussein and aided the Shi'ite uprising at the end of the war. After suppressing the uprising, Saddam resumed the drainage project with a vengeance, not so much to create agricultural land but to destroy the Marsh Arabs. The "Third River," renamed the "Saddam River," later paralleled by the "Mother of Battles River" and the "Loyalty to the Leader River," diverted the Euphrates and helped to drain the southern marshes, while the "Prosperity River" captured Tigris tributaries and drained marshes to the north and east. These projects further reduced the total wetland area by 90 percent by the year 2000. About 300,000 Marsh Arabs became refugees, and UN Resolution 688 called for humanitarian assistance for them. Much of the land became barren in the worst example of rapid desertification in the Near East in recent times, and wildlife declined disastrously: the otter became extinct, and more than forty species of waterbirds and migratory birds were endangered. In Iraq generally, there are no significant protected areas for ecosystems including birds and other wildlife. This is unfortunate because Iraq is an important part of the migratory flyway between Eurasia and Africa. Spawning grounds for fish, including seventy-one native species and twenty-one introduced species, important to the Persian Gulf fisheries, declined as a result of drainage of the marshes. Since 2003, local people have begun reflooding some areas of the marshes on their own, but the effort necessary for restoration, if that is what will be done, will be expensive and the time will be long. Restoration may not be fully possible because the volume of the two rivers is continually reduced by dams and irrigation upstream.

The presence of petroleum was noted in Mesopotamia in very early times. There were seeps of it, and although no one used it as fuel, they did find minor medical and other applications. Alexander the Great performed an infamous experiment with it, setting fire to a boy named Stephanus, who almost died as a result. Exploitation of petroleum in Iraq began after World War I. The first successful well was bored by a Turkish company near Kirkuk in 1927. By 1939, before World War II intervened, France was receiving from Iraq half of all the crude oil it needed through pipelines that reached the Mediterranean Sea in Tripoli, Lebanon, and Haifa, Palestine. In 1960 Iraq joined the Organization of Petroleum Exporting Countries (OPEC) and was receiving 61 percent of its governmental revenues from oil royalties. Between 1973 and 1978 its oil revenues

Toward the end of the first Gulf War, when Iraq's defeat was inevitable, Saddam Hussein ordered his retreating forces to set fire to Kuwaiti oil wells, resulting in immense damage to the environment. (Peter Turnley/Corbis)

increased by a factor of more than thirteen. The oil industry was fully national-
ized by 1980.

Iraq's geopolitical importance as one of the world's most important sources
of oil contributed to the political and military embroilments of the years around
the turn of the century. In August 1990 Iraq invaded and attempted to annex the
small but oil-rich country of Kuwait, and as a result the UN, with the encouragement of the United States, imposed an embargo on oil exports by Iraq. A U.S.-led
invasion of Iraq, supported by many other nations, reestablished an independent
Kuwait. The war had many negative environmental effects caused by explosions
and the movement of military vehicles across the landscape, but by far the most
damaging impacts came from the deliberate opening of oil wells by the Iraqis,
which they then set on fire. At one time more than six hundred oil wells were
burning, turning day to night and resulting in horrendous pollution of air, land,
and sea. The total amount of pollutants generated by the fires is estimated at
500,000 metric tons per day, including sulfur dioxide, carbon monoxide, soot including metal-contaminated dust particles, and combustion products such as
benzopyrene, PAHes, and dioxins, of which the last three mentioned are carcino-
genic. The fires lasted several months, although not always at the level just described, and the smoke plume extended for hundreds of kilometers. Soot and un-
burned oil contaminated the surface of the sea, along with flows of oil from
uncapped wells, killing at least thirty thousand seabirds and unknown numbers
of fish, with possible effects on the seafloor. The environmental damage in
Kuwait alone is estimated at $40 billion. A UN compensation fund was estab-
lished, in part to pay compensation for losses including environmental damage.

In 1995 UN Security Council Resolution 986 allowed the partial resump-
tion of Iraq's oil exports, with the proceeds to be used to buy food and medicine
(the "Oil for Food" program). Securing a healthy environment would assist in
meeting basic humanitarian needs. But in 1998 Iraq ended its cooperation with
UN inspectors, and a U.S.-U.K. bombing campaign began. The second Gulf War
began in March 2003, and U.S.-led coalition forces soon occupied major oil
fields and key sites in Baghdad, including the Oil Ministry. Some oil fires were
started by Iraqi forces during the invasion, but the environmental damage from
that source, at least, was not nearly as great as in the first Gulf War. It should be
noted, of course, that ordinary peacetime operation of oil fields and refineries
has its environmental costs, and that these are likely to be greater in the period
of restoration of oil production, pipelines, and shipping, especially with damage
and sabotage by insurgent operations that are occurring at present.

The environmental damage of warfare and its aftermath has been serious and
complex in Iraq. Due to the collapse of sewage treatment facilities, huge quantities of raw sewage and industrial waste are discharged into rivers every day, espe-

cially into the Tigris at Baghdad. Water treatment systems are damaged, subject to leaks, and poorly maintained, exposing the population to waterborne illnesses including typhoid, dysentery, cholera, and polio. Groundwater is subject to pollution by oil spillages. Urban solid waste is not being removed due to destruction of equipment and unavailability of parts and repair. The problem is complicated by military debris, including toxic and radioactive material such as depleted uranium. The United States fired over 290 metric tons of depleted uranium projectiles during the first Gulf War, and an undetermined but probably greater amount during the second Gulf War. U.S.-U.K. attacks on suspected chemical, biological, and nuclear facilities during and after the first Gulf War may have released some hazardous materials into the environment. During the time of its operations in Iraq, UNSCOM destroyed large quantities of chemical agents and weapons, but this was done with extreme care to avoid damage to the environment.

Large tracts of forest and date-palm groves were leveled and burnt during the Iran-Iraq War and the Gulf Wars, whether deliberately as part of military preparations or inadvertently during conflict. The war also caused a halt in spraying to treat groves of date palms and citrus that had been attacked by an aphid-like insect called *dubas* and other insects, including the white jasmine fly, a pest accidentally introduced from Southeast Asia in 2000. In consequence, production of fruit dropped and prices rose disastrously.

Institutions important for study of environmental problems and provision of expertise for their solution suffered massive physical damage and at present are barely functional. These include government agencies, schools and universities, scientific corporations, libraries, and museums. The Federation of Arab Scientific Research Councils was located in Baghdad before the war. The reestablishment of these institutions must be a concern of those responsible for Iraq's future. This is, in fact, one of the recommendations of a UNEP study issued in 2003 after the occupation of Iraq by the U.S.-led coalition. The others include surveying the environment and locating environmental "hot spots," cleanup and restoration of water supply and sanitation systems, making sure that environmental protection is integrated into the postconflict reconstruction process, and international cooperation to address the chronic environmental problems facing Iraq.

CONCLUSION

What is the future of the Mediterranean environment? This may not be a question that environmental history can answer directly, but to aid in answering it is certainly one of the purposes of studying the relationships of human societies to the natural environment of the Mediterranean in the past. It was also the subject of a study by the Mediterranean Action Plan and United Nations Environment Programme, including a large number of experts from nations of the Mediterranean area, published in 1989 (Grenon and Batisse). This report concentrated on the coastal strip, with its daunting problems of land-use planning, urban management, and pollution control, as the most crucial area where the future of the Mediterranean environment hangs in the balance. As the editors cautioned, "protection of the terrestrial and marine coastal strip will be very difficult in the long run because of growing human pressures and the vulnerability of its natural environment" (viii).

If present-day trends were simply projected into the future, the picture would look very bleak indeed. That is because most of the factors causing environmental decline are either increasing or show no signs of abatement. Population and economic consumption continue to rise, while poverty and landlessness have resisted attempts to reduce them. To be realistic, bad economic policies, shortsighted politics, corruption, and illegal trade continue to be endemic to the region. War, terrorism, and other forms of violence have if anything increased in the last few years; one need only mention Palestine and Israel, the nations of the former Yugoslavia, and the nearby Persian Gulf region that is so intimately connected with the Mediterranean. It would be a mistake, of course, to think that these problems are limited to the countries just named. Considering all these factors, then, it is necessary to understand the negative trends, but also to look for positive trends that could reverse some of the processes of decay and render the future more hopeful.

A major worldwide problem affecting the Mediterranean that is expected to increase is climatic change due to global warming. Most scientists believe that this phenomenon is intensified by the production and release of "greenhouse gases" such as carbon dioxide, methane, and nitrous oxide by industry and other human activities. They expect that the warm desert conditions now characteristic of the margins of the Sahara will continue to move northward. This is due not just to an increase in average temperatures but also to the expected movement of weather patterns such as storm tracks to the north. Over most of the Mediterranean region, average rainfall is predicted to decrease as concentrations of greenhouse gases rise. As P. H. Gleick summarized it, "Future climatic changes effectively make obsolete all our old assumptions about the behavior of water supply. Perhaps the greatest certainty is that the future will not look like the past. We may not know precisely what it will look like, but changes are coming" (Gleick 1993, 138).

Desertification of significant areas in the Mediterranean region is a distinct possibility, and it is the subject of studies being sponsored by the European

Union. Another result of global warming is the rise of mean sea level around the world due to the expansion of the volume of the oceans and the melting of continental ice in Greenland, Antarctica, and glaciers elsewhere. In the Mediterranean, a higher sea level would present a threat of watery invasion and salinization to low-lying coasts and river mouths, nowhere more severely than the Nile Delta in Egypt, which lies close to the level of the Mediterranean, is the area that produces the great majority of Egypt's crops, and is home to most of Egypt's people. Any solution to the problem of global warming would involve the cooperation of nations around the world. In the Kyoto Protocol of 1997, the world's major industrial nations agreed to reduce their emissions of greenhouse gases by various percentages by 2012. It should be noted that such industrializing nations as China, India, and Brazil did not accept reductions, and that the United States withdrew from the agreement in 2001. The Mediterranean nations that belong to the European Union (most of those that are on the northern margin of the sea) agreed to a reduction of 8 percent but at this writing had not made necessary progress toward that goal. Global warming is, of course, a problem without a local solution.

Acid precipitation is another atmospheric problem that, however, is more closely related to the emission of pollutants into the atmosphere within a region, although the Mediterranean is also affected by entry of these pollutants from outside the region. The northern and western Mediterranean nations and Israel are highly industrialized, and the rest of those on the south and east are striving to become so as rapidly as possible, threatening the continued increase of acid rain. Acids in rain can kill marine organisms in lakes and other small bodies of water, can cause agricultural damage, and are a factor in the death of trees and the reduction of forest cover. In the late 1980s, the number of coniferous trees in European Mediterranean nations suffering from defoliation was determined to be over 60 percent in Greece, 30 percent in Spain and Yugoslavia, 20 percent in France, and 5 percent in Italy and Portugal. The discrepancy in these figures probably represents differing standards of measurement, but there is no doubt that the problem is serious and increasing.

One of the most important driving forces of environmental damage is population growth. As Russell King remarked, "Demography lies at the heart of Mediterranean destiny" (1990, 164). It is a problem in the Mediterranean basin as a whole, but there is significant difference between the north (e.g., Spain, France, Greece, Italy, and the former Yugoslavian republics) and the south (Algeria, Egypt, Libya, Morocco, Syria, Tunisia, and Turkey). In the north, the rate of population increase in the past few decades has fallen to the point of stability, and some countries are actually experiencing a decline that is expected to continue. In the south, rates of increase have also declined, but not to anything like the decline in the north. This is crucial, because this is the part of the basin with the greatest problems of water and food supply. The southern Mediterranean countries added 65 million people between the years 1985 and 2000—to 226 million—and are expected to increase by another 100 million between the years 2000 and 2025. If this increase occurs, by 2025 the population of the Mediterranean area will be just over 500 million, with two-thirds living in the south. In 1950 the total was 150 million, with two-thirds living in the north. Programs aimed at reducing the rate of population growth exist in several of the nations in the south and have met with some success. Although the availability of contraception is certainly an important factor in lowering the rate of population growth, the most effective long-term trends in this respect in places that have achieved such a decline have been amelioration of poverty; economic security for the aging segment of society; education, particularly of women; and a rise in the status of women.

Urbanization presents a host of related problems that must be faced in the future. About 80 percent of the population will live in cities in 2025, a figure already reached in some northern countries. Environmental quality in the Mediterranean basin will depend more on the way urbanization occurs than on any other factor that is amenable to planning. National and municipal governments need to provide for safe water supply, recycling, treatment of sewage, and elimination of toxic waste. Air pollution can be combated by energy conservation, movement to nonpolluting sources such as solar technology in this proverbially sunny region or geothermal energy generation such as the plant constructed in Tuscany by the Italians, and the reduction of motor vehicle pollution at the source. Comprehensive planning to preserve historic city centers, provide for green space and tree planting, and control suburban sprawl is a necessity, not merely a desideratum.

Food dependency and potential food shortages will worsen in the south unless agricultural production can be increased. Sadly, intensification of agriculture in recent history has been connected with soil, water, and air pollution from fertilizers and pesticides and from mechanization, all of which are expected to increase. Some of these activities involve discharges into the sea, which are expected to increase and must be lessened or prevented wherever possible. Irrigation has increased salinization, as in Egypt, and has involved the depletion of nonrenewable underground sources of aquifer water, as in Libya. Desalination of seawater, although requiring energy and relatively difficult in the exceptionally salty Mediterranean, as well as improved drainage, could help. Irrigation dams need to be combined with revegetation projects in the watersheds that feed them, to reduce erosion and salinity. Preservation of native Mediterranean seed stocks is important to enable the recovery and development of varieties adapted to local conditions.

The objective in fisheries should be careful studies aimed at understanding fish populations and establishing a permissible level of catches for sustained use of this renewable resource, applied throughout the almost landlocked sea. The alternative is the near loss of fisheries. Aquaculture is possible, but it must be regulated to avoid the presently increasing problem of nutrient pollution that produces algal blooms and other forms of marine biotic destruction, sometimes exhausting the available oxygen from the water and killing virtually every form of life in the affected environment.

Industrial development will undoubtedly increase during the coming decades and will occupy land, coastlines, and ecologically sensitive areas; dangerous forms of pollution will continue unless carefully planned and regulated. Unfortunately this is difficult because governments see industry as a positive force in economic development, and industry has political and financial muscle to wield everywhere in the decision-making process. The creation of industrial parks with facilities for pollution control put into place as development occurs would lessen some of the problems. Clean production, reducing the amounts of pollution, would go far to help the environment, as long as that means actual technological measures to improve industrial processes and not just "green" industry as a public relations ploy. Independent inspections and legislation requiring antipollution equipment are certainly essential basic measures. Finally, oil transportation must absolutely be made safer and pollution-free.

Tourism is likely to increase exponentially in the future in this region, one of the world's most popular destinations, and almost all the nations in the region are eager to develop this major source of trade surpluses. The nations with the largest number of tourists and the largest amounts of income from tourists are France, Italy, and Spain. For several nations, including Malta, Cyprus, and Egypt, tourism is the leading economic activity, or was before recent disruptions of the industry. There are two negative factors likely to prevent its growth in the years to come. One is crime directed against tourists, including terrorism, which at this writing has almost shut down the tourist industry in some countries. It is possible to envision war, terrorism, and all forms of violence decreasing and almost disappearing as in some happy times of the past, but it is not possible to judge the likelihood of such a blessing. Still, there are parts of the Mediterranean that seem calm and remain popular with tourists, and most Mediterranean people have a remarkably resilient friendliness and hospitality. The second negative factor is environmental decay, since it is the Mediterranean environment that attracts tourists in the first place. Beaches have been closed by incidents involving sewage and algae, along with the hypertrophy of native and introduced species of seaweed. Swimmers have caught diseases while swimming, including diarrhea, dermatitis, conjunctivitis, cholera, and typhoid, and as a result the tourist industry has constructed huge water recreation centers in places where the seawater is too unpleasant for bathing. Few people enjoy viewing congested areas or breathing industrial smoke or urban smog. As fish become rarer, the price of a seafood menu has risen above many tourist budgets. Tourism itself, if allowed unregulated growth, can destroy the very places that created it. The annual number of tourists, around 184 million (although their lengths of stay vary and are usually short, mostly in the summer), represents an addition to the resident population and its impacts. Coasts completely occupied by hotels, outdoor restaurants cheek by jowl, beaches carpeted by beach chairs, and must-see attractions with lines blocks long eventually repel even the less discriminating traveler. The more discriminating traveler may require ecotourism, and that is even more dependent on having something that at least looks like nature. The word gets around, and tourists plan trips to less polluted places outside the Mediterranean. There is, in other words, another potentially economically rewarding aspect of environmental protection.

The future of the Mediterranean forest will almost necessarily be that of the protection of watershed and the prevention of erosion. This will require sustained effort, since studies predict that 50 percent of the northern forests and 100 percent of those in the south will have disappeared between 1990 and 2025, a result that needs to be prevented. Significant portions should be protected as national parks and wildlife habitat in which nonconsumptive human use can continue. In some limited areas, they can be used for production of timber on a strictly sustained-yield basis. But the amount of forest that remains is far below the desirable area, so further deforestation cannot be permitted. Substitutes should be found for fuelwood in places where peasants still deplete the forests to heat their houses and cook their meals. For example, Cypriots, Spaniards, and Israelis often heat their water by solar power, an idea that could find many applications elsewhere and would save trees. Forest should be reestablished, especially in mountainous areas that are presently bare and support nothing but emaciated, ravenous goats. Instead, native wildlife can be considered for reintroduction and the ecosystem gradually restored. In particular, the once-great forests of the cedars of Lebanon should be fostered and extended in the mountains of that country, in which the image of a cedar graces the national flag. International cooperation through the United Nations and other associations can be of crucial importance in achieving some of these goals, but the most important requirements for an environmentally livable future are understanding and determination on the part of the people of the Mediterranean region, and lasting peace.

References

Attenborough, David. 1987. *The First Eden: The Mediterranean World and Man*. Boston and Toronto: Little, Brown.

Brandt, C. Jane, and John B. Thornes, eds. 1996. *Mediterranean Desertification and Land Use*. Chichester: John Wiley and Sons.

Carrington, Richard. 1971. *The Mediterranean: Cradle of Western Culture*. New York: Viking Press.

Faulkner, Hazel, and Alan Hill. 1997. "Forests, Soils, and the Threat of Desertification." In *The Mediterranean: Environment and Society*, eds. Russell King, Lindsay Proudfoot, and Bernard Smith, 252–272. London: Arnold.

Geeson, N. A., C. J. Brandt, and J. B. Thornes, eds. 2002. *Mediterranean Desertification: A Mosaic of Processes and Responses*. Chichester, UK: John Wiley and Sons.

Gleick. P. H. 1993. *Water in Crisis: A Guide to the World's Fresh Water Resources*. Oxford: Oxford University Press.

Gomaa, Salwa Sharawi. 1995. *Environmental Threats in Egypt: Perceptions and Actions*. Papers in Social Science 17, Monograph 4. Cairo: American University in Cairo Press.

Grenon, Michel, and Michel Batisse, eds. 1989. *Futures for the Mediterranean Basin: The Blue Plan*. Mediterranean Action Plan and United Nations Environment Programme. Oxford: Oxford University Press.

Harrison, Paul, and Fred Pearce. 2000. *AAAS Atlas of Population and Environment*. Berkeley and Los Angeles: University of California Press.

King, Russell. 1990. "The Mediterranean: An Environment at Risk." *Geographical Viewpoint* 18: 5–31.

McNeill, John R. 2000. *Something New under the Sun: An Environmental History of the Twentieth-Century World*. New York: W. W. Norton and Company.

Pastor, Xavier. 1991. *The Mediterranean* (*Greenpeace: The Seas of Europe*). London: Collins and Brown.

Proudfoot, Lindsay, and Bernard Smith. 1997. "From the Past to the Future of the Mediterranean." In *The Mediterranean: Environment and Society*, eds. Russell King, Lindsay Proudfoot, and Bernard Smith, 300–305. London: Arnold.

Thornes, John. 2001. "Environmental Crises in the Mediterranean." In *Geography, Environment, and Development in the Mediterranean*, eds. Russell King, Paolo de Mas, and Jan Mansvelt Beck. Brighton, UK: Sussex Academic Press.

United Nations Environment Programme (UNEP). 1992. *World Atlas of Desertification*. London: Edward Arnold.

7

CASE STUDIES

In a picture dominated by localism, case studies are extremely important.

—JOHN R. MCNEILL, *THE MOUNTAINS OF THE MEDITERRANEAN WORLD*, 1992

CASE STUDY A: ENVIRONMENTAL PROBLEMS OF CITIES IN MESOPOTAMIA

Climb up on to the wall of Uruk, inspect its foundation terrace, and examine well the brickwork; see if it is not of burnt bricks; and did not the seven wise men lay these foundations? One third of the whole is city, one third is garden, and one third is field, with the precinct of the goddess Ishtar. These parts and the precinct are all Uruk.

—*EPIC OF GILGAMESH*

In the study of the history of Mesopotamia, scholars of the early and mid-twentieth century concentrated on subjects connected with the elite segments of society, such as literature, law, religion, and astronomy. But recent investigations have turned the attention of historians to what were formerly considered to be peripheral subjects, such as environmental, biological, and botanical studies; aspects of subsistence and artifact production and exchange; and materials science research. This broadening of focus has made possible a more holistic approach. Today civilization can be understood more fully because it is seen in its larger setting and in its interactions with the environment.

The appearance of the city as a mode of human relationship to the natural environment established a pattern that would increase in importance for the rest of history. The characteristics of civilization—the state with its religious and political institutions, the specialization of human occupations, the stratification of society into social classes, and the development of arts such as monumental

184

Part of the clay brick wall of the ancient Sumerian city of Uruk, displayed in the Pergamon Museum, Berlin, is decorated with colored terra-cotta cones. It may serve as a metaphor for the separation of the city from its rural and natural environs. (Photo courtesy of J. Donald Hughes)

architecture, writing, and the measurement of space and time—appeared first and developed most fully in these densely populated human centers. These centers first emerged around 3000 BC in Mesopotamia. Early cities were not large by modern standards: Ur, a dominant city of ancient Sumeria (the southern part of Mesopotamia), attained a population of perhaps thirty-seven thousand. Another great city of the region, Uruk, covered 250 hectares (almost a square mile) and may have accommodated forty thousand people. For a human aggregation of even this size, it was necessary for agriculture to develop to the point where the labor of a farm family could produce enough food to feed not only itself but others as well. This happened with the invention of the ox-pulled plow and incipient irrigation. It was also necessary for society to create the institutions that would organize food production and distribution, the import of useful materials, and the defense of one city against the appropriation of its lands and goods by another. As cities increased their consumptive capacities and the specialization of tasks, a major aspect of their economic relationship with their hinterlands came to be one in which they exchanged manufactured products for raw materials, including the various materials necessary for manufacture.

The Mesopotamian Environment

In order to gain meaningful comprehension of Sumeria and the subsequent cultures, such as the Babylonians and Assyrians, that inhabited Mesopotamia, it is essential to have some understanding of the climate and environment of this region, since these factors are the foundations on which Mesopotamian civilization was built. How much the ancient landscape differed from its condition in modern times must also be investigated. The role of the Tigris and Euphrates, the two major rivers that bring water to Mesopotamia and drain it, is paramount. Except for some montane margins and the Mediterranean coastal sections, this region would be desert were it not for these rivers and the irrigation they make possible. The salt and silt carried down from the surrounding mountains by the rivers created the landscape within which the area's inhabitants lived. The water of the rivers, diverted into artificial canals, was the basis of the people's subsistence.

The fertile, sandy soil of Mesopotamia was easily turned by the ox-drawn plow, but Mesopotamia received little rainfall; indeed, the average precipitation in modern times is less than 200 millimeters (8 inches) in the plain from the Persian Gulf northward to about latitude 36 degrees. Though rivers provided the needed water, their flow was so undependable that control by major irrigation works was required. The most extensive and labor-consuming achievement of the Mesopotamians, therefore, was the system of canals that brought water to the fields. These irrigation works wrested rich provisions from the basic fertility of the land. Thus a Mesopotamian king could feel justified in listing the excavation of a new canal, along with the conquest of an enemy city, as the major events of his reign. Once the infrastructure of water delivery was constructed, a new agriculture enabled a much larger human population to live in expanded settlements, and many people no longer had to work on the land, so that specialized occupations could flourish in the cities.

Almost all of the early cities arose in the flat, alluvial land of Mesopotamia, where stone or metal for building rarely occurred, and there were few trees large enough for construction, so clay was the dominant material used in ordinary structures. Urban dwellers erected mighty works of baked and unbaked clay bricks: palaces, thick city walls, and temples rising on lofty ziggurats. Pottery vessels were manufactured and used in the production, storage, and consumption of beer, wine, oil, milk, and milk by-products.

The Bronze Age

Copper and bronze metallurgy appeared at much the same time as the early cities, and the period (approximately 3000–1000 BC) is therefore called the Bronze Age. Some of the earliest metal objects were formed from copper and copper alloys that occur naturally in metallic form around the Near East. In the search for additional supplies, metalworkers turned to smelting copper oxide ores, adapting techniques from pottery firing to obtain the high heat that was required. In the process, it was discovered that alloys of copper were superior to the pure metal in hardness and the ability to keep an edge. The preferred metallic material proved to be an alloy of copper and tin called bronze. Metallic ores were almost completely lacking in southern Mesopotamia, but a considerable range of metal artifacts, including jewelry, tools, and weaponry, were manufactured and traded in that region. Evidence exists in the cuneiform and archaeological records for various metalworking and metallurgical activities, including extractive processes. Copper, tin, silver, lead, iron, gold, and their ores were imported from territories on the margin of the Tigris-Euphrates valley and at even greater distances. Commercial, military, and diplomatic efforts were applied to acquire metals. Tin was rare and had to be imported over long distances. The source of tin may have been in alluvial deposits of cassiterite from several places abroad, including the Mediterranean lands, the Indus Valley, and Bactria (modern Afghanistan). Household utensils, decorative and ceremonial objects, and armor and weapons came to be made of bronze.

Religion and Attitudes to Nature

The Mesopotamians were polytheistic. Their gods were awesome powers, visualized in human shape, who had created the world and who continued to direct it from their homes in the sky and beneath the earth. Many gods were personifications of celestial objects, such as Utu, the sun, and his father, Nanna, the moon. The air, waters, and vegetation had their patron deities. These and many other gods and goddesses were worshiped in rectangular temples built of clay brick and adorned with stone and wood, raised above city streets on one-story platforms, and then even higher on the step pyramids called ziggurats. Images of gods and their worshipers, carved in stone with prominent noses and huge round eyes, inhabited temple interiors.

The urban attitude, then, is in large part a desire to improve the environment by imposing order on chaos, and by so doing, to control the natural world. This is done through making direct physical changes. It is also attempted through religion, which in the early urban period was done mainly through sacrifice. Sacrifice is the presentation of gifts to the gods in the form of valuable objects; products of the fields such as grains, oil, or wine; animals domestic or, less commonly, wild; and human sex, human blood, or human lives, in the hope that the gods will reciprocate. Sacrificial worship is directed at contacting the forces that animate the universe, feeding them so the environment will be abundant, satisfying their desires so they will not bring disasters, and in general controlling them for human benefit.

Priests not only indicated the times of planting and harvest and required that a proportion of each year's crop be given for their own support and the worship of the gods, but also created a need for products from far away. The roofs of ever-larger temples required long, straight timbers that the treeless Mesopotamian plain could not supply, and sculpted images had to be made from stone that the alluvial soil did not contain. The far-ranging merchants obtained these products. Merchants were an important segment of Sumerian society, but it must not be imagined that they represented free enterprise. The rulers managed their activities, and when they traveled to other cities, their status was that of quasi-ambassadors. These first merchant-venturers traveled by land and sea. To the east, they traded along all the coastlands of the Persian Gulf and even as far as the Indus Valley. In the west, they brought fine woods from Lebanon and copper from Cyprus, and they were in touch with Egypt almost continuously by way of the Red Sea. Fine timber was also obtained from the Zagros and Taurus mountain ranges, both of which suffered the beginnings of a long process of deforestation. Thus the Mesopotamians transformed distant landscapes through trade.

Apparently, writing first arose in dealings between priests and the merchants who supplied their needs. The earliest examples of symbols impressed into clay are small tokens marked with conventional pictographs, used to keep track of animals, foodstuffs, and other products conveyed from one place to another or kept in temple precincts. Early in the Sumerian period, similar pictographs began to be inscribed on small clay tablets. Soon scribes adopted a triangular-pointed stylus as better suited to the sticky clay surface, and the pictures evolved into arrangements of wedge-shaped (cuneiform) marks. Since these stylized pictures could be given abstract meanings and/or phonetic values, they came to represent the full range of language. Letters, laws, contracts, poems, historical and other documents—all could be written in cuneiform. Sumerian cuneiform was adapted to other languages spoken in Mesopotamia and beyond.

Urban people in Mesopotamia displayed a new attitude toward the natural environment, almost diametrically opposite to that of the hunter-gatherers, herders, and early farmers. A stance of confrontation replaced the earlier approach of reverence and cooperation, as if the barrier of city walls and the rectilinear pattern of canals had divided urban human beings from nature outside them. This attitude can be traced in literature from Sumerian times down through Akkadian and Assyrian writings, which often use the image of battle to describe the new relationship.

In Mesopotamian cities, gender roles seem to have been somewhat strictly divided, and male warriors tended to fill those that were dominant. Men wrote most of the literature, too, although almost certainly not all of it, and images of combat and conquest involving warrior heroes are prominent as a result. Mesopotamian literature is complex, however, and includes survivals from pre-urban mythologies; warrior goddesses and male earth gods also appear in the sagas of cities. Still, it seems likely that attitudes to nature might have been more supportive if women had constituted a governing force in urban societies.

In creation myths, a hero-god (Enlil or Marduk), the embodiment of civilization, confronted and defeated Nature in the form of a female monster of chaos (Kur or Tiamat). In this outlook, the natural chaotic state of the universe could be overcome and order established only through the conquests made by the gods. The constant labor of their human followers created the order of the city, with its straight streets and strong walls, and the regular pattern of canals in the countryside, all of which were believed to constitute an earthly imitation of the heavenly order that the gods had established. The renowned Mesopotamian law codes included provisions to protect the ownership and divisions of the land.

Careful observation of the heavens motivated many of the Mesopotamian inventions that most influence modern civilization. Records were kept of the movements of the sun, moon, planets, and stars, and the zodiac was divided into the familiar twelve constellations. The circle was divided into 360 degrees, the hour into 60 minutes, and the day into 24 hours. Mathematics also became a systematic aid to land management, business, and the legal system.

The *Epic of Gilgamesh*, whose earliest documents come from Sumeria, and which is perhaps the oldest long poem of which we have any knowledge, reveals the urban Mesopotamian sense of the distinction between the tame and the wild, between civilization and wilderness, and shows a new and hitherto unfamiliar attitude of hostility toward untamed nature. Enkidu, the hairy man of the wild, first appears in the poem as a friend and protector of beasts, but he is a nuisance and even a menace because of that, releasing animals from the traps of hunters and warning them away from their ambushes. When a woman tamed him and introduced him to the pleasures of bread, wine, and sex, his former animal friends caught the scent of civilization upon him and fled. Entering the city of Uruk, he met and struggled with King Gilgamesh, who after winning the almost equal battle adopted him as his inseparable companion. Together they went on a quest to distant mountains for cedarwood. The cedar forest happened to be a sacred grove protected by the wild giant Humbaba, and his defeat and death at the hands of the two heroes became a symbol for the subjugation of the wilderness by the city. Gilgamesh promptly ordered the cedars cut down and carted to Uruk for use in building his new palace. Humankind's proper endeavor with wild things, in the Mesopotamian view, was to tame them. Native animals such as the onager and water buffalo were added to the sheep, goats, pigs, and cattle already domesticated by their ancestors. Animals that successfully resisted subjugation were hunted mercilessly; the epic says that Gilgamesh killed lions simply because he saw them "glorying in life," whereas he as a human being was all too conscious of his mortality.

The stories of Gilgamesh reveal the existence of two warrior institutions under the king: the council of elders or experienced warriors, whose survival through numerous battles gave them the right to be heard by the king; and the assembly of all warriors, who had the right to voice their opinion on declarations of war. The king was expected to consult these bodies but was not bound by their decisions. Though the king was in theory "first among equals," his power could not be doubted. All warriors were citizens with full political rights. In Sumerian cities, there were only two legal classes: citizens and slaves, and the slaves were not especially numerous.

Post-Sumerian History

Sargon the Great, king of Semitic Akkad in central Mesopotamia, conquered the Sumerians around 2300 BC, subordinating the formerly independent governments of the Sumerian city-states to the rule of an empire. Respecting the gods of conquered peoples and claiming divinity himself, Sargon and his successor kings controlled Mesopotamia for two hundred years, controlling major decisions on water use and the allocation of other resources. One of his noted accomplishments was an expedition to the cedar forest (probably in the Amanus Range) in the northwest, doubtless a raid to gain access to timber that was floated down the Euphrates River to the heartland of the empire. His grandson, Naram-Sin, repeated the cedar forest foray but made it more permanent, leaving a garrison and establishing a fort to safeguard the timber road. A celebrated stele with Naram-Sin's image marking the extent of his conquest was left in what is now eastern Turkey.

Eventually the Sargonid Dynasty, weakened by barbarian invasions and local uprisings, gave way and was eventually replaced by the Third Dynasty of Ur, another empire but this time a Sumerian one whose ruler claimed the title "King of Sumer and Akkad." Its most memorable king, Ur-Nammu, promulgated a law code with notably mild punishments. He ruled through provincial governors chosen from among Sumerians and Akkadians alike, and in the city-states he posted representatives called *ensis*, who were usually chosen from the local people of the particular city. His son Shulgi collected fair and well-recorded taxes in goods and animals. Factories controlled by the government produced textiles, flax oil, leather, flour, beer, and wine, and provided jobs for thousands of women and men. Despite such an enlightened policy, the Third Dynasty succumbed to secession and foreign invasion.

Hammurapi, king of the great city of Babylon, united Mesopotamia by conquest soon after 1800 BC. He is deservedly remembered for his systematic law code, which survives almost intact in a stone stele carved with a cuneiform text that reveals much about the social and economic structure of his kingdom. It safeguarded the economy by protecting businesspeople, traders, and landowners. The leading warriors, priests, and civil servants made up the nobility, which had special privileges, particularly in the ownership and management of land. All other free citizens were commoners, whose status also was guarded by law. Many of these citizens conducted the long-distance trade that supplied both luxuries and essential raw materials; they operated under the law's protection. The third and lowest class in this stratified society was that of slaves, who were recognized as human beings but whose limited status depended upon their value as the property of others.

Assyria, expanding from a homeland on the banks of the Tigris River in northern Mesopotamia, became one of the ancient world's largest and most powerful empires, dominating all of Mesopotamia including Babylon, the entire seacoast of the Mediterranean Levant, and briefly Egypt and much of Iran. Assyria appeared as an important kingdom in the thirteenth century BC, made its most important conquests in the ninth and eighth centuries BC, and perished at the hands of an alliance of enemies in the late seventh century BC. The first impression one gains from the surviving ancient sources is that Assyria was a militaristic, aggressive, and merciless power. This is in part due to the fact that many of the sources were written by Assyria's adversaries and victims. For example, the prophet Isaiah, who lived in Jerusalem in the eighth century BC and witnessed an Assyrian invasion, spoke for his God, Yahweh, as follows (Isaiah 10:5–6):

> *The Assyrian! He is the rod I wield in my anger,*
> *The staff in the hand of my wrath.*
> *I send him against a godless nation,*
> *I bid him march against a people who rouse my fury,*
> *To pillage and plunder at will.*
> *To trample them down like mud in the street.*
> *But this man's purpose is lawless,*
> *And lawless are the plans in his mind;*
> *For his thought is only to destroy*
> *And to wipe out nation after nation.*

This violent impression is borne out by the inscriptions of the Assyrian kings themselves. From their point of view, it was their own god Assur who had ordered them, his earthly representatives, to conquer. Assurnasirpal II (ruled 883–859) announced, "I am merciless . . . first in war, king of the world . . . who has trampled down all who were not submissive to me." He carried out these words with deliberate atrociousness. He demanded booty and tribute as tokens of submission, and he was not the only Assyrian king to use mass deportations of subject peoples as instruments of his economic policy. Some of them were used in the construction of his magnificent palace in his new capital of Kalah (Nimrud), and others labored at specialized tasks in military support, industry, trade by land and sea, and agriculture. Assyrian kings also appropriated the gods of their conquered peoples, most notably those of Babylon, and carried their images to Assyrian temples or appropriated them in situ in the temples of dependent cities.

The Assyrians put together the world's first truly efficient Iron Age army, a military machine of unequalled efficiency. Their artisans were excellent metallurgists, skilled in smelting, enameling, and inlaying and in making iron arms and weapons, including swords, shields, helmets, long lances, and heavy bows. Slingers, bowmen, and lancers composed their infantry, and they also could field a formidable cavalry and chariots drawn by horses. Assyrian engineers could construct bridges, tunnels, moats, and efficient siege weapons that could take fortified towns, as their reliefs show with vivacity. Many of the Assyrian army units were levied from conquered territories and placed under Assyrian commanders. Supplies for the army, including fuel for iron smelting and timber for the construction of siege machinery, placed severe demands on the forests located in the mountains to the north and east of the Assyrian plains.

The effect of Assyrian conquests on the Near Eastern landscape was major, but not completely destructive. Systematic mass deportations of agricultural peoples must inevitably have affected the lands for which they had been caring. These lands were parceled out as rewards to Assyrian military men and businessmen, making them great landowners, but as they consolidated their properties the kings began to feel threatened by them and took measures to break up the landholdings. This led to the so-called Great Revolt of 827–822 BC by the rural nobility and powerful provincial governors against King Shalmaneser III, which was put down with extreme difficulty and great cruelty by the court and army. Subsequently, administrative reforms strengthened royal power, weakened the large private landholders, and established new, smaller provinces that limited the resources available to the governors and reduced their ability to threaten the central government.

The Assyrians were great city-builders, and the urban centers of the kingdom grew as the empire was able to commandeer the resources of the lands it dominated. Sennacherib (ruled 750–681) moved the capital to the renowned city of Nineveh, which he fortified with 1,500 watchtowers and five walls that were wide enough on top for three chariots to drive abreast. The walls were pierced by 15 gates and protected by three moats, one inside the other. A great aqueduct supplied the city with water.

One must balance the portrayal of the Assyrians as a ruthless militaristic people with the evidence of a highly developed civilization that was found in a city such as Nineveh. There were thirty temples. The royal palace had a huge park that contained the world's first botanical and zoological gardens and a sacred fishpond. Indeed, the fishpond served as the city's identifying cuneiform sign. Nineveh was not only the administrative center of the empire but also an admirable focal point of literature and learning. Assurbanipal (ruled 668–627) and his brother Shamash-shum-ukin (668–648) assiduously collected clay cuneiform tablets from throughout their dominions and created the Great Library of Nineveh, which contained works on mathematics, astronomy, and astrology as well as official correspondence, business records, dictionaries, and many texts of literary value. The discovery of the remains of these records by Austin Henry Layard in 1845 made possible the recovery of much of Sumerian, Akkadian, and Assyrian literature, including major fragments of the *Epic of Gilgamesh*. Thus we can be grateful for the culture preserved by Assurbanipal, even if he did hang the head of the rebellious king of Elam in his garden so he could gloat over it.

It is interesting that the book of Ezekiel in the Hebrew Bible, written between 593 and 571 BC after the fall of the Assyrian Empire, describes the rise and fall of Assyria through an environmental image. Assyria, states Ezekiel, was like a cedar of Lebanon, the greatest of trees in the world, and the peoples of its empire found shelter like the birds and animals of the forest in its branches and under its shade, sharing the springs of water that nourished it. But foreigners had ruthlessly cut down the tree, and the peoples of the Earth left because they could no longer find shade under its branches, while the birds and wild animals gathered on its fallen trunk, seeking refuge in vain, since the springs of water had dried up. This view of Assyria as a devastated ecosystem is remarkably sympathetic, coming as it does from a member of one of the nations formerly threatened by Assyria. It was the resurgent Babylon, however, one of the nations that had laid Nineveh waste, that had also carried Ezekiel's nation into captivity. Ezekiel simply believed that Assyria, like other nations, was an example of God's care for the nations in endowing them with physical and biological resources. It is certainly likely that the image of the downfall of Assyria as a fallen tree may reflect the extent to which Assyria overused the natural resources of its empire.

Environmental Problems of Mesopotamia

Throughout their history, the Mesopotamian cities were in continual danger of flooding; the Tigris and Euphrates rivers rose over their banks unpredictably, destroying settlements and fields, which is one reason why the inhabitants regarded nature as chaotic. The system of canals and dikes was intended to serve not only for irrigation but also for flood control. Cities accumulated mounds, rising above the plain and, if they were lucky, above the floods.

Growth in number and density of population produced problems of pollution, waste disposal, and the spread of diseases, affecting the health, stature, and longevity of the inhabitants. Drinking water was drawn from wells, rivers, and canals subject to contamination. Mesopotamian documents mention the danger of death from drinking bad water. To pollution from sewage and offal were added wastes from industrial activities such as metallurgy, leather tanning, and pottery kilns. These accumulated until rain washed them into rivers and groundwater. A few early cities arranged for removal or built sewers and latrines similar to those found in the ruins of Knossos on Crete. Wastes as well as the concentration of human bodies and stored foodstuffs attracted opportunistic organisms.

Human health suffered by every measure. Neolithic villagers were less healthy than hunters and herders, but city dwellers showed further decline; studies of their skeletal remains show that they were shorter in stature, lived briefer lives, suffered more from bad teeth and bones, and were subject to communicable diseases. To these dangers must be added warfare, slavery, and human sacrifice. An unconscious trade-off had been made that forfeited quality of life for quantity of humans and more security for the community. For individuals, urban life was not necessarily an improvement over earlier societies.

The mud and silt carried by rivers and canals settled out rapidly. Continuous dredging was required to keep the canals flowing, and the excess material piled up along their banks until the canals were fifteen meters or more above the surrounding fields, so that they could no longer serve to drain the land and presented a danger of overflow in time of flood. Salinization, the accumulation of salts in the soil as a result of water evaporation, is a constant menace wherever irrigation is practiced in dry climates, and in Mesopotamia it was unfortunately widespread. The lower sections of the valley of the Tigris and Euphrates have, as noted earlier, an arid climate, and the soils are relatively impermeable, two factors that can lead to an accumulation of salt at levels that are harmful to soil texture and to crops, and which if continued can eventually make agriculture impossible in the areas affected. Irrigation water carried into low-lying areas was allowed to evaporate, and over the years in this land of low humidity and meager rain, the salts accumulated. The conditions of poor drainage also made it difficult to correct the situation by leaching salt from the fields. Groundwater became increasingly saline. Farmers tried to adapt to the changing conditions by planting salt-tolerant barley instead of wheat, which was more sensitive to salt. Crop yields declined; for example, in the agricultural lands of the city of Girsu, there was a reduction of 47 percent in production of grain in liters per hectare between 2400 and 2100 BC. In extreme cases, cultivated plants were entirely incapable of growth in salinized soil. Because of this, many areas had to be abandoned, while new sections were brought under irrigation and cultivation until they in turn suffered similar effects. A survey by Thorkild Jacobsen and Robert Adams (1958) found evidence in temple records of increased salinity and declining yields in southern Mesopotamia between 3500 and 1700 BC, and they identified this trend as contributing to the breakup of Sumerian civilization. Their judgment: "That growing soil salinity played an important part in the breakup of Sumerian civilization seems beyond question" (1252).

They considered that measures to counteract siltation and salinization might have succeeded, but that the authorities in charge of the cities were preoccupied with military adventures and political intrigues. Regrettably, the same problem persists in modern times, exacerbated by conflicts over the development of some of the world's richest deposits of petroleum.

Conclusion

The recovery of evidence concerning the relationship between ancient Mesopotamian civilizations and the natural environment depends on the preservation of clay tablets bearing cuneiform texts and other artifacts that represent economic activities, including artistic representations of these activities and the vessels in which various products were contained. These artifacts were collected in careful archaeological excavations conducted by European, American, and Iraqi scholars over many decades of the nineteenth and twentieth centuries. The objects themselves and the records relating to their excavation were placed in museums such as the Louvre, the British Museum, the Oriental Institute of the University of Chicago, and the University of Pennsylvania Museum in Philadelphia. Most important, they were located in the National Museum of Iraq and in regional museums in Iraq. This was especially true of material excavated recently, since Iraq understandably wished to keep treasures representing its historic heritage within the country, and to stop the removal of these treasures to museums in other nations. Unfortunately, during the first Gulf War of 1991, nine out of thirteen of the regional museums experienced looting and lost valuable antiquities. This loss, although serious, pales in comparison to the looting and vandalism of the National Museum of Iraq and other archaeological sites in 2003 and the destruction of museum records and others in the National Library. The resulting loss to the study of the history of this land and people, who produced the first civilization, the first writing, and many other contributions to literature and science, is a crime against human knowledge.

Isolated mounds in a desert environment now mark the locations of the renowned cities of ancient Mesopotamia, and photographs taken from space show that the fertile land occupies a fraction of its former extent. These effects are not the result alone of climatic change or of warfare, although both have occurred throughout the centuries. They embody the epitome of ecological disaster caused by human actions. In Mesopotamia, perhaps more than in any other region, there is a clear relationship between environmental degradation brought about by destructive human actions, whether intended or not, and by cultural decline.

CASE STUDY B: ECOLOGY AND THE DECLINE OF THE ROMAN EMPIRE

*All the cities that have ever held dominion or have been the
splendid jewels of empires belonging to others—some day men will
ask where they were. And they will be swept away by various kinds
of destruction: some will be ruined by wars, others will be destroyed
by idleness and a peace that ends in sloth, or by luxury, the bane of
those of great wealth.*

—SENECA, *MORAL EPISTLES*

*Our fields turn mean and stingy, underfed,
And so today the farmer shakes his head,
More and more often sighing that his work,
The labor of his hands, has come to naught.
When he compares the present to the past,
The past was better, infinitely so.*

—LUCRETIUS, *THE WAY THINGS ARE*
(*DE NATURA RERUM*)

The problem of the economic, political, and military decline of the Roman Empire, from the apex of its power and geographical extent in the first and second centuries AD to a state of fragmentation and depression five hundred years later, has intrigued many historians because it does not seem to have a simple explanation. Indeed, it invites many theories of causation.

The history of civilizations can be read as an illustration of the fact that human events do not take place outside the context of the natural environment. Human societies have natural ecosystems as their setting, and cultures are in large measure adaptations of human groups to the ecosystems within which they live. The health of the first depends on the health of the second. Civilizations flourish only as long as the ecosystems they depend upon. Nowhere is this clearer than in the Mediterranean basin and its environs, where many early civilized societies had their origin and ran their course. Many of these societies suffered fates that seem to have visible relationship to environmental factors. Much of the "Fertile Crescent" is desiccated and poor today, North African cities have been buried in the desert sands, and once-thriving ports now lie under strata of erosional sediments. Thus, the question is reasonable: How far was environmental deterioration responsible for the decline of ancient communities, and to what extent were the actions of human beings instrumental in degrading the natural environment and causing the collapse of their own societies?

Human beings have always made changes in the natural environment; it is impossible to live without doing so. But some changes allow a functioning ecosystem to continue in balance with the human societies that depend on it, while changes of a different kind or magnitude hamper and abrade the natural systems that support all living things. Changes of the first kind make human life possible and pleasant for extended periods of time, whereas changes of the second kind will eventually render human life unpleasant, difficult, and less sustainable. The history of ancient civilizations provides us with examples of both kinds of interactions.

In the year 1776, Edward Gibbon presented the first copy of a new book to King George III. It was *The Decline and Fall of the Roman Empire,* now considered one of the monuments of English prose historical writing. In it, he gave several reasons for the fall of Rome: two of them were that the bureaucracy and the military were overgrown, causing the vast empire to collapse of its own weight, and that the flower of Christian manhood was flocking into monasteries, depriving the civil powers of their aid. Other reasons have been offered, then and now. Even while it was happening, there were those who maintained that the moral decadence of contemporary Romans, or their failure to worship the gods, or the one God, was weakening society. Later theorists blamed overextension of territory, decline of population, class struggle, the lack of individual initiative, the drain of precious metals to the East, lead poisoning, soil exhaustion, and changing climate.

Each of these explanations has been vigorously championed, and some of them are supported by good evidence. What seems certain to unbiased observers of the debate over the decline of Rome is that no single factor is likely to have been the cause of such a complex phenomenon. It is therefore necessary to look for a series of contributing causes. It was not just one thing that brought down Rome, but a number of processes that interacted. One of these, quite probably, was the Roman mistreatment of the natural environment, including overexploitation of scarce natural resources such as forests, and failure to find sustainable ways to interact with the ecosystems of Italy and the Mediterranean.

Among the elements in any adequate theory of multiple causation are various environmental processes. The history of Roman civilization can be interpreted as an illustration of the general principle that human history takes place within the sphere of the natural environment and is in large measure subject to the laws of ecology. The health of human societies depends on the health of the ecosystems on which they depend.

Study of archaeological reports, scientific studies of deposits of silt from erosion and ancient pollen grains, and ancient writings and inscriptions has led me, and others, to the conclusion that environmental factors were important causes of the decay of Roman economy and society—though not the only causes—and that the most important of these factors were produced by human activities. The result of the process of deterioration is also evident in the landscape, where the fact of environmental degradation in the Mediterranean basin is quite clear. Olive presses of Roman date have been found in desert areas of Tunisia, where today there are no trees at all, much less olives, in sight. The once-flourishing cedar forests of Lebanon are represented today by a few small groves. And the sediments of rapid erosion still exist and can be studied in lowlands not far from Rome and along coastlines all around the inland sea.

Roman art reveals an ambivalent attitude toward the environment. The Ara Pacis (Altar of Peace) from the time of Augustus portrays Italia as an earth-goddess surrounded by all the tokens of agricultural peace and prosperity, including herd animals and the plentiful fruits of harvest. But a fresco from Pompeii shows goats grazing freely in a grove around a statue of Artemis, the protector of the wild whose sacred space is obviously being violated. Reliefs on the imposing monument, Trajan's column, show Roman soldiers hacking their way through forested Dacia, chopping down trees, and carrying away logs to be built into siege terraces, catapults, and battering rams, or stacked into huge piles to be set on fire as beacons.

Some of the wisest Romans were aware that humans often abuse the natural world. The Roman statesman, dramatist, and philosopher Seneca remarked, "If we evaluate the benefits of nature by the depravity of those who misuse them, there is nothing we have received that does not hurt us. You will find nothing, even of obvious usefulness, such that it does not change over into its opposite through man's fault" (*Quaestiones Naturales* 5.18.15). Environmental problems are, in this view, the passionless revenge of the Earth on those who fail, through ignorance or avarice, to practice the art of the attentive guardians of the land.

Chapter 2 describes many ways in which the Romans wasted resources and degraded their environment. Among these are deforestation, overgrazing, erosion, salinization, exhaustion of the soil, devastation by war, depletion of animals and plants, pollution, and failure of public health measures. Since people are not irrational enough to want to reduce their quality of life and undermine the structures that meet their needs, it may be asked why the Romans failed to find an environmentally sustainable path. Again, there is no single reason, but several considerations can be suggested that might lead toward an adequate answer.

Knowledge of the Ways of Nature

The ways in which a society relates to nature are shaped by the knowledge of nature that it possesses. People who live close to the natural world as gatherers, hunters, herders, and agriculturalists accumulate a considerable body of knowledge through trial and error. The ancestors of the Romans imparted this knowledge to them in the form of myths, traditions, and commonsense injunctions. But such lore is full of inaccuracies and misunderstandings. Agriculture could be carried on fairly well through use of tried-and-true methods as long as it was not disrupted by natural disasters, political and economic exactions, or war, but these disruptions, as already noted, were all too common. Some of the Greeks had been particularly interested in learning what makes the world of nature work. They asked questions that we now regard as ecological. Aristotle and his brilliant student, Theophrastus, were two who did so, although none of their followers advanced beyond them. Reason was doomed to failure in this endeavor to the extent that it was unsupported by observation and unchecked by experiment. Even keen minds like those of the Greeks Democritus and Epicurus and the Roman Lucretius, all of whom advanced the atomic theory, engaged in doctrinaire speculation rather than empirical research into the phenomena of nature. The ancients barely achieved science worthy of the name. This was particularly true of the Romans, who admired and patronized Greek science but made few theoretical advances of their own. Pliny the Elder, the most noted Roman naturalist, was a collector and compiler of information true and false, not an independent thinker. His *Natural History* is an amazing collection of unusual facts and fictions, some from his own observations but most from earlier writers such as Theophrastus. Science did not advance far enough among the Romans to enable a sound theoretical understanding of the web of life. Roman agricultural writers made a number of useful comments that bear on ecological subjects, but they can hardly be called students of ecology. A good practical grasp of what needs to be done to achieve a sustainable return existed in agriculture and, perhaps to a lesser extent, in pastoralism and forestry. But economic, political, and military factors intervened to prevent the achievement of a trial-and-error modus vivendi with the Earth. In particular, a balance with nature is a condition of peace and is easily upset by war, and war all too often raged across the lands of Italy and the empire during the Roman centuries, devastating the land, killing those who cared for the soil, and disrupting the orderly movement of food supplies.

Roman Technology

In order to be able to find a sustainable balance with nature, a society must have an appropriate technology. One is tempted to say that the Roman technology was relatively appropriate, since it was simple, utilizing human and animal power for the most part. But because the Romans brought their efforts to bear

The Roman emperor Trajan ordered the construction of this column of marble, 35 meters (115 feet) high, to celebrate his conquest of the forested land of Dacia (modern Romania). Its reliefs show military operations, including the cutting and use of timber. (Photo courtesy of J. Donald Hughes)

cumulatively over centuries, much total damage was done. Even relatively simple technologies can be destructive. The Roman dependence on wood and charcoal for energy, which resulted in major inroads into the forests, demonstrates this very kind of cumulative damage.

Ironically, the very technological achievements of the Romans that moderns admire most in retrospect, such as aqueducts, sewers, roads, and bridges, are those that show most clearly their ability to alter and control nature in ways that were sometimes productive, but often destructive. The aqueducts do seem estimable, and the Romans were deservedly proud of them. The sewers promoted waste disposal and public health in the city but polluted the river and created dangers to health downstream, since there was no treatment of the effluent. The roads, many of them straight as an arrow, were long enough that if the major highways were joined together in a line, they would stretch 90,000 kilometers (56,000 miles), more than enough to circle the world twice, and if smaller roads were added, 322,000 kilometers (200,000 miles), giving a total distance almost long enough to reach the moon (Hopkins 1988, 759–760). Plutarch described their construction: "The roads were driven through the countryside, exactly in a straight line, partly paved with hewn stone, and partly laid with impacted gravel. Gullies were filled in, [and] intersecting torrents and ravines were bridged, so that the layout of the road on both sides was the same, and the whole work looked level and beautiful" (*Gaius Gracchus* 7). With the well-constructed, aesthetically pleasing bridges that carried them over rivers and gorges, such as the famous bridge at Alcántara, Spain, carrying a road 47.5 meters (156 feet) above the Tagus River on six arches, the roads in this vast system improved communication and enabled the exploitation of natural resources at great distances. They encouraged development of agriculture, mining, and industry farther from the metropolitan center by providing access to more distant areas. Because of them, more forests were felled, and plants and animals transported, with the result that they were introduced to new lands or extirpated in their original ranges. Roads increased human mobility and reduced the inaccessibility of marginal territories, amplifying the impact of the Romans on the natural environment. Unfortunately they also encouraged the spread of malaria because mosquitoes bred in the pits and ditches that had been created during construction and subsequently became filled with rainwater.

Granted the proficiency of Roman engineers, it is strange that they seem to have halted on the verge of further steps, such as the harnessing of water and wind power, for which the Greek inventors Ctesibius and Hero of Alexandria had already created the basic theoretical models. Even steam power might have been within their grasp. The revolution that those technological advances might have made possible remained unachieved for unknown reasons, although a slave economy, psychological resistance, a desire to preserve jobs, and failure to develop interchangeable parts have all been suggested as reasons for the failure. The Roman Empire possessed the requisite natural resources, especially metals, in abundance. Perhaps the most convincing reason for the failure of the Alexandrians and their Roman successors to utilize their inventions fully is their lack of a tradition of investing in innovative technology. There was venture capital for the import business, but not for untested new machines of dubious application. One can imagine Sestius, a substantial grain and wine merchant, being asked to invest his fortune in a little wind-driven pinwheel or a whistling steam kettle, neither of which did anything more than go round and round. Yet it would be pointless to criticize the Romans for not achieving the Industrial Revolution fifteen hundred years before it actually occurred. If they had developed their technology further without improving their attitudes toward and knowledge of nature, their impact on the natural environment would have been swifter and more destructive.

Civilization and Social Control

Ability to interact positively with the environment also requires a degree of social organization and control. This is necessary because the attainment of goals that are desirable for the society as a whole entails sacrifices, even if they are small ones, on the part of individuals. For example, the owner of a herd of goats will not refrain from grazing his animals on a hillside where small trees are growing just because it would be good to have a forest there in a few years. He puts a marginal gain for himself ahead of a larger gain for his community. So the state must exercise some degree of coercion. But too much coercion, ostensibly for the public good, can be extremely harmful to the environment if it is directed to exploitation and not informed by ecological understanding.

The Roman government established policies affecting agriculture, forestry, mining, commerce, and so on, but citizens were allowed tolerable latitude of choice within these guidelines. Huge works like aqueducts, canals, and roads show that the republic, and later the empire, had ways of getting cooperation, but it can well be imagined that the common people evaded regulations if it seemed in their interest to do so. Roman citizens had carefully defined duties to the state, and the state concerned itself with the use of various categories of land. The Roman Empire attempted to interfere in and control the lives of its citizens to a greater extent in the days of Diocletian and afterwards. The emperor's autocracy could not approach the ability of a modern state with computer technology to keep informed about its citizens and be sure that they performed their social duties. Notwithstanding that, it was able to do a lot. The environmental result of the considerable degree of social control exercised by the Romans can be seen in their ability to build huge public works as well as to reach out and use resources located at great distances. Roman roads and ships brought timber from the Alps and Lebanon, grain from Egypt, and tin from beyond the Pillars of Hercules.

A damaging aspect of Roman social organization as it affected the environment was its direction toward war. A balance with nature is a condition of peace. Rome was a warrior-dominated society, never at peace for very long, and its army exercised strong political influence throughout its history. Even the famous *Pax Romana*, which lasted, with a few breaks, for two hundred years, did not end wars along the frontiers, and was followed by fifty years of war in the third century AD that left no major province untouched by campaigns and battles. War exacted a debilitating toll from agriculture, because military campaigns devastated the countryside, slaughtered farmers and their families, and requisitioned or destroyed crops and buildings. Farmers were conscripted and had to spend time fighting instead of caring for the land, so terraces and irrigation works were left neglected. The mere passage of armies living off the country and trampling the crops was a calamity, but calculated "environmental warfare," in which an enemy's natural resources and food supplies were demolished, was not uncommon either.

The Social Structure of the Roman Empire

The economy of the Roman Empire was organized primarily to benefit the upper strata of society: the landlords rather than the peasantry, the rich rather than the poor, the masters and certainly not the slaves. Above all, power and economic benefit were concentrated in the office of the emperor and in his household. The emperor controlled a separate treasury, and his edicts had the force of law. The emperor could affect the economic structure and thus environmental impact, although it must be noted that Roman understanding of economic policy was limited and that nothing like environmental policy existed.

A word about the social standing of women is appropriate at the outset. The Roman tradition was extremely patriarchal in the sense that a woman's status was determined by that of the man to whom she was primarily related: her father before her marriage, her husband during marriage, her son during widowhood, and a related male guardian in the absence of the first three. She would be a member of the class to which the man belonged but could not exercise direct political or legal participation. A woman might be of citizen class, but she was not a citizen. For example, there were no female senators or magistrates. There was often an empress (*augusta*), but only in relationship to a male emperor. At the same time, upper-class women could wield considerable political and economic power through or on behalf of their husbands or other related men. Their influence in terms of environmental impact was, therefore, very considerable. Whenever the masculine pronoun is encountered in what follows, therefore, this fact should be remembered.

The Roman class structure began in the single city-state of Rome and had spread across an empire that surrounded the Mediterranean Sea during the centuries before Augustus. The upper classes (*honestiores*) were designated in terms of both family and wealth. The three upper classes were the senators, equites, and decurions.

Highest among these were the senators, originally the descendants of patrician families, but augmented by worthy and wealthy plebeians and, by imperial times, powerful colonial and provincial leaders. Qualification as a senator required property in the form of landed estate. The influence of this class in terms of environmental impacts was paramount, since its members owned a major fraction of the land and along with the emperor had a determining role in legislation. Many of them owned extensive ranches (*latifundia*) where cattle, sheep, and pigs were raised and various agricultural goods—including oil, wine, grains, and cloth—were produced. They might invest in various enterprises, but social and sometimes legal barriers kept them from direct involvement in trade and commerce.

The second class consisted of equites (equestrians), who were required to be freeborn (the emperor Tiberius added the qualification of two generations of free birth) and to have property. This class included an aristocratic Roman core, leading men from the colonies, and businessmen (*negotiatores*). Its members held important public offices, including the prefectures in charge of supplying the cities with food and other necessities. The principal tax collectors (publicans) were equites. This class was the leading force in trade and commerce, organizing markets and investing in shipping, and controlling trade in such products as grain, wine, and oil. Crucial to the treatment of the environment, equites formed consortiums (*societates*) to control the exploitation of important state-owned natural resources such as mines, quarries, and forests (Rostovtzeff 1957, 110).

The third class, the decurions, consisted of members of municipal councils in the empire. Like the first two classes, they were expected to share high birth, wealth, and good reputation. They also performed public duties, including local administration, finances, and tax collection. Environmental decisions on the local level, such as land questions or the creation of infrastructure, often fell to them. These three classes had an advantageous status under law. Together they constituted less than 2 percent of the population of the Roman Empire, but they controlled the vast preponderance of its economy.

There was a steady concentration of landownership into the hands of the wealthy, because over time the small farmers had gone into debt to the large landowners, placing their land in security, and had not been able to pay back the loans (Simkhovitch 1921, 104). "For neither knowledge nor effort can be of any use to any person whatsoever, without those expenses which the operations require" (Columella, *De Re Rustica* 1.1). The landowners controlled huge estates where tenant sharecroppers and contingents of cattle-herding slaves directed by bailiffs performed the labor. Large consortia of wealthy families contracted with the government to exploit mines, forests, and other resources. There were also many smaller enterprises run by citizens of lesser rank, including freedmen, and these often depended on funding from the more affluent. Trade was usually conducted by the dependent enterprises, since commercial occupations were considered demeaning to those of the upper classes, although nothing prevented the latter from profiting through the labor of the former. They therefore provided the construction funds and other venture capital for merchant fleets that crisscrossed the Mediterranean and other nearby seas, including the near Atlantic Ocean, the Black Sea, and the avenue through the Red Sea to the Indian Ocean. The more direct route to India had been opened by a Greek sailor named Hippalus using the monsoon winds not long before the rule of Augustus. Pliny the Elder (first century) complained that Rome spent a hundred million sesterces every year in trade with India, and this money was exchanged not for bulky resources that might have benefited the common people, but for more easily transported luxuries such as spices, jewels, fine woods, birds, and delicate textiles to satisfy the tastes of the elite. A statuette of the Indian goddess Lakshmi was found in Pompeii, providing one bit of evidence of this trade (Charlesworth 1951). Roman exports in exchange included gold, silver and silverware, works of art, wine, pottery, and slaves.

The lower classes (*humiliores*) were also divided into three recognized groups, the freeborn poor, freedmen, and slaves. The freeborn poor were those who had never been slaves and had no slave ancestry. The majority worked on the land, making up the preponderance of the population of the Roman Empire, but some were self-employed tradesmen, others were skilled or unskilled workers who sold their labor, and a large group made up a portion of the urban unemployed. Although it is not well represented in historical sources, the effect of this class on environmental change must have been major, taking place in a multitude of everyday decisions and activities throughout the empire.

Freedmen were emancipated slaves and those descended from them. They were eligible for Roman citizenship, but that citizenship might carry certain disabilities, and like the freeborn poor they might be dependent on patrons from among the *honestiores*. Some became rich, powerful, and envied, like the bureaucrats appointed by the emperor Claudius (41–54) or the millionaire Trimalchio, who although fictitious undoubtedly represented a class that existed. According to Petronius's *Satyricon* (first century), Trimalchio was considering the purchase of the entire island of Sicily to add to his landholdings, an obvious exaggeration but one with an edge of truth. Most of the freedmen were not rich, but people economically akin to the freeborn poor.

The vast majority of the Roman population was rural, lived at a subsistence level, and led precarious economic lives (Garnsey and Saller 1982, 28). Most of them, probably 80 percent, labored on the land. Small farmers managed only a slight surplus margin of production, if any, and that was quickly commandeered by the landlords and by imperial taxes, which weighed heavily on the agricultural sector throughout the period with which we are concerned. The government's basic concern, above all, was to finance the imperial court, the bureaucracy, and the military, and to construct public buildings and infrastructure such as roads and sewers. The danger of famine in the countryside, and consequent political unrest, was recognized by the progressive emperors Nerva (96–98) and Trajan (98–117), whose edicts required landowners receiving government grants to make subsistence payments to poor children in rural districts. The practice of such government handouts did not genuinely ameliorate the socioeconomic structure that was in effect designed to benefit the social elite. The distributions may have been directed toward preventing unrest among the poor, and possibly also toward increasing the population.

No picture of the effect of the Roman economy on the natural environment can be complete without an investigation of the role of slavery. As Aldo Schiavone recently explained it, "The use of slaves became the ideal functional means of agricultural exploitation, slave labor the basis of all manufacturing, and the owner of land and slaves the ultimate protagonist of every organization of production. . . . [I]t is impossible to separate the society of Rome—its material foundations, obviously, but also its ideas, convictions, mentality, ethics, and even its anthropology—from the context of slavery" (Schiavone 2002, 122–123). The status of a slave was equivocal; the law treated a slave as a "speaking tool" (*instrumentum vocale*). Mine slaves demonstrate what is so often pointed out: the Romans had slaves to compensate for a shortage and inefficiency of machinery. But gradually through Roman history, especially during the second century, a degree of personhood was recognized. A freed slave could become a Roman citizen, although of lower class. Slaves made up at least 35 percent of the population of Roman Italy during the time of Augustus, and were similarly numerous elsewhere in the Roman Empire. Slave owners among the higher classes were known to possess hundreds or thousands. In AD 61, for example, Lucius Pedanius Secundus had four hundred house slaves; field slaves and mine slaves were much more numerous. Many agricultural slaves worked on latifundia, mostly assigned to outdoor jobs that often required them to sleep outdoors where they were exposed to malaria. A typical assignment for a slave was to watch grazing animals and take them to mountain pastures in summer, but Columella (*De Re Rustica* 3.3.8) tells of specialized slaves skilled in viticulture, and there were slaves who managed business for their owners. Slaves could not be expected to exhibit initiative in improving agricultural methods or ensuring sustained yield of renewable resources because they were obliged to obey the commands of their owners to do mechanical and repetitive work, had little time at their own disposal, and if they produced agricultural surplus or other income, it went to their masters. They lacked even the marginal incentives that tenant sharecroppers had in planning for personal and family rewards. A slave economy cannot as a rule benefit the environment: slave labor enabled exploiters to do more damage, and a slave class whose members were liable to be sold anywhere could not establish a relationship of responsibility with the land. Slaves performed most of the actual work in forestry, such as felling trees (Aubert 2001, 101). In addition, the fact that, as Pliny the Elder expressed it, "agricultural operations are performed by slaves with fettered ankles and by the hands of malefactors with branded faces" (*Natural History* 18.4.21) increased the tendency of Roman citizens to think of farming not as care of the earth but as degrading work. The Roman system of slave labor was not only corrupting of human values but environmentally destructive as well.

Roman Ideas about the Natural Environment

It may seem that the prevailing ideas and attitudes of a people about nature ought to influence their actions in regard to nature. The Romans had a typical practicality that led them to look at the physical world primarily in terms of its value to them. The statesman Cicero, for example, commended the cunning of human hands and the many things they had accomplished, including agriculture, domestication, construction, mining, forestry, navigation, and hydrology, and concluded with an intriguing notion: "Finally, by means of our hands we endeavor to create as it were a second world within the world of nature" (*De Natura Deorum* 2.60). The world, Seneca noted, was designed for human use. Metals, for example, are hidden deep in the earth, but humankind possesses the ability to find them and dig them out. Well-planned projects make the Earth more beautiful and serviceable. Beauty and function are one and the same, he maintained. Domestication actually improves plants and animals. Expanding civilization remedies a defect of the wilderness, which was formerly only a stomping ground of wild beasts or a desert waste. Not all human efforts were seen as enhancements of nature, however. Cicero must have been in a different mood when he declared, "the products of nature are better than those of art" (*De Natura Deorum* 2.34). Other writers observed that the results of human activities were at times neither beautiful nor useful. The poet Horace derided "the owner contemptuous of the land" (*Carmina* 3.1.36–37).

Pragmatic Roman attitudes could promote wise use of resources, with an eye to sustained returns in the future. But they could also encourage those who sought short-term gain at the expense of sustainability. At few times was there an attitude that would have encouraged a vigorous attempt to guard the integrity of natural ecosystems and to live in balance with them. The Romans had conquered the political world; they seem to have thought they had conquered nature as well. Here can be seen an important root of ecological crisis. To use nature as a slave by right of conquest, without considering her ability to meet the demands placed upon her, was shortsighted practicality indeed.

The love of the Romans for their native land in Italy is well-known; they thought of their own city and its surrounding countryside as the world's central place, particularly blessed by the gods. Virgil addressed his country with the images of the ancient agricultural religion: "Hail, great mother of harvests! O land of Saturn, hail! Mother of men!" (*Georgics* 2.173–174). In the time of the monarchy and early republic, Roman attitudes toward the natural environment were tinged with deep respect and often fear of uncanny forces they sensed in nature. Their gods were *numina*, mysterious presences in the natural world, invisible and rarely depicted in human form. They were regarded as operating in spheres such as agriculture, and as protecting nature from major human injury. Any natural phenomenon could be seen as the result of a god's action. Therefore any human activity affecting the environment could be seen as attracting the interest of or provoking the reaction of some god or goddess, and ought to be undertaken with caution. How much this attitude moderated Roman practicality and actually protected nature can only be guessed. The same religion that established protections for nature also provided ways to gain exceptions from those protections. For example, Cato the Elder advised his readers to use a generic prayer beginning "Whether you be god or goddess who inhabits this grove" whenever they wished to cut down trees in a sacred grove where such an action was forbidden (*De Agricultura* 139–140). The prayer was to be accompanied by the sacrifice of a pig. Religious prescriptions are not designed to support ecological values, although they may happen to function in that way, but all too often become mere ritual prohibitions.

Roman religion had a strongly agricultural flavor, reflecting the early observances of farm families close to the land who depended on the annual cycles of nature for subsistence. The numerous Roman religious festivals followed a calendar based on the round of activities of the ancestral farm, from the hanging of the plow on the boundary marker in the celebration of Compitalia in January to the holidays dedicated to Saturn and Bona Dea, deities of the soil, in December. Roman gods associated with the natural environment were extremely numerous, from great deities like Mercury, god of flocks, to local spirits of particular springs, like Juturna. The Romans honored gods of the farmhouse and storehouse (*penates*) and of the fields (*lares*), which were at the same time symbols of the powers of nature. A god or goddess was the growing spirit of every major crop, such as Ceres of grain or Liber of wine. Beyond that, every major and minor activity of the farm had a *numen* that could be invoked for its success, such as Vervactor for first plowing, Repacator for second plowing, Imporcitor for harrowing, Insitor for sowing, and even Sterculus for manuring. On the margins of the farm lurked Silvanus and other wild forest gods.

Pliny the Elder complained that people abuse Mother Earth (*Mater Terra*). She was an important goddess, but it seemed to many Romans that Earth was growing less fertile and less able to sustain human beings (the philosopher Lucretius believed this to be a natural process of aging). But others, like Columella (*De Re Rustica*, preface 1–3), differed, holding that environmental deterioration is due to human failures. Earth is not growing old, he maintained: the blame for her infertility lies in lack of human care; declining crops are our fault, not hers. Earth gives her gifts to those who treat her well and punishes thoughtless farmers or greedy ones who try to extract from her what she is not willing or perhaps able to provide. Environmental problems are, in his view, the passionless revenge of the Earth on those who fail, through ignorance or avarice, to be attentive guardians of the land. The responsibility for environmental balance, therefore, rested in the hands of humankind.

Roman Imperial Economic History and the Natural Environment

Augustus inaugurated a period of peace within the Roman Empire, although wars of conquest and defense continued on the frontiers. The *Pax Romana*, which lasted with few interruptions until the early third century, enabled an expansion both of population and of the economy. Piracy at sea and banditry on land were suppressed, opening golden opportunities for commerce (Finley 1999, 156). Expansion of the economy, however, meant that the richer orders of Roman society, which constituted a tiny minority, became richer. Augustus consolidated Rome's domination of the landscape of the entire Mediterranean world and beyond it in western Europe, Claudius added Britain in the first century, and Trajan conquered Dacia (Romania) in the second century and conducted a campaign (ultimately unsuccessful) in Mesopotamia (Iraq). This territorial control enabled Roman officials to enrich themselves at the expense of provincials, and Roman entrepreneurs to exploit natural resources over an area of some 7.8 million square kilometers. The demand for timber in construction of large buildings and ships caused deforestation over large tracts of land.

Economic output increased due to an increase in scale rather than in efficiency. In agriculture, new areas were cleared and brought into production, and labor-intensive methods increased food production, a necessity during this period of rising population, especially because a larger proportion of people were living in cities. Rome, the capital, grew in population and area, spreading over the surrounding countryside and blurring the distinction between city and suburbs

(Morley 1996, 85). The imperial government continued to make every possible effort to provide a subsidized grain import to feed the urban poor, and under Augustus eighty thousand tons of grain were distributed annually free to 200,000 people (Schiavone 2002, 96). Augustus treated Egypt, one of the major sources of wheat imports, as his own personal property, thus securing the role of the emperor in providing the food supply. The emperor also took interest in an organized business that provided a continuous supply of gladiators and animals for entertainment in the amphitheater. These two enterprises constituted the famous "bread and circuses" (*panis et circenses*), intended to keep the mob happy and prevent revolution. They were also notorious engines of environmental depletion: the grain dole drained the agricultural production of the provinces, and the arena contributed to the extirpation of large species of wildlife everywhere.

Depletion of natural resources and wasteful methods of exploitation were underlying causes of the crisis of the third century, which manifested itself in the form of shortages and ruinous inflation. There were no important advances in agricultural or industrial technology, except possibly the use of the vertical undershot waterwheel to grind grain. A population decline probably began with the great plague of 166–182 under Marcus Aurelius (emperor 161–180), after which violence, food shortages, and returns of the plague kept recovery at bay. Some emperors attempted state control of resources, raising taxes including those collected in kind, and assuming direct control over some aspects of trade. Archaeological deposits of amphorae, for instance, reveal that the imperial bureaucracy took charge of supplying oil and wine from Spain to Rome during the reign of Septimius Severus (193–211), but he was the only emperor of the first half of the third century to exercise consistent control of such matters (Macmullen 1988, 14). The emperor Caracalla (211–217) enlarged the Roman citizen body and the tax rolls by an edict granting Roman citizenship to all free men in the empire in 212.

The size of the army, and its consumption of finances and resources, increased by leaps and bounds. Commanders in the provinces sought to seize power, plunging the empire into fifty years of sporadic warfare. The battlefields were in the settled regions of the central empire, and destruction was visited upon houses, barns, orchards, and the rural population. The average period of rule of the emperors between 235 and 284 was two years, hardly enough time to establish policy, and all of the emperors were military men, few of whom had any understanding of principles of economy. Prices rose astronomically; the price of wheat in Egypt, for example, was 8 drachmas per artaba in the second century, 24 drachmas in the mid-third century, and 220–300 drachmas in the late third century (Duncan-Jones 1990, 147). Emperors facing financial emergency increased the minting of coinage, thus exposing it to debasement. Due to the inflation of the value of precious metals, the cost of the metal in coins rose above their face value, forcing the issuing of coins in less valuable materials such as bronze or lead (perhaps with an easily eroded wash of silver). Silver could be used for coins of higher denominations in a never-ending inflationary process. But the monetary economy is an artificial one; if it is to operate successfully, it must reflect the underlying natural economy. The tax base of the empire, which depended on agricultural productivity, was shrinking. There were further onslaughts of plague in 251–266 and afterwards, and emperors made up a deficit of manpower by allowing groups of barbarians to settle within the empire.

Population decline continued to be a problem in later antiquity, meaning fewer farmworkers, so that reductions in population and agricultural production tended to be synergistic. This exacerbated what was happening at the end of the third century: although constant warfare and periodic plagues were also to blame, there was a chronic agricultural decline deriving from the environmental damage the Romans had caused. Food was becoming scarcer, prices were rising, and there was a general shortage of labor. Diocletian attacked these problems vigorously, if not entirely successfully. His edicts on occupations required civil and military officials, decurions, landowners, and shipowners to provide heirs for their positions, namely their own natural or adopted sons. This was intended to counteract the drop in population by encouraging those in essential jobs to produce children. It also turned these professions into hereditary castes. The same principle was later applied to others on whom the food supply depended: bakers, butchers, and food merchants, and then craftsmen, postal employees, workers in state factories, and ordinary soldiers. It was no surprise, then, when agricultural workers were included. Laws helped landlords tie peasants to the soil, restricting their freedom of movement and requiring them to remain permanently attached to the latifundia, thus beginning a process that led toward eventual serfdom in a later period. The flight from the land was largely stopped, but at the price of individual freedom.

Diocletian enacted edicts against inflation while restructuring the empire to guarantee central control and restrict local autonomy. The Edict on Prices, issued in 301, set maximum allowable charges for various commodities, services, and wages. It lists approximately one thousand specific items in an attempt to control profiteering. The categories listed include food, raw materials such as timber, clothing, transportation, service charges, and wages. The exhaustive catalog reveals some of the environmental impacts of the Romans at the time; for example, prices are given for wild game such as gazelles, pheasants, and sea urchins, and for furs including badger, leopard, and sealskin. The prices overall appear to be fair, although perhaps a small percentage under the rates actually prevailing at the time. If Diocletian's wages and prices had become the actual practice, ordinary people would have had a supportable standard of living (Williams 1985, 131). The edict did not succeed, however, because it failed to take account of such economic principles as wholesale and retail, supply and demand, and the availability of natural resources. When the price set by the edict for a commodity was considerably below what the market demanded, the item disappeared from the shelves in stores; as Lactantius wrote, "Men were afraid to display anything for sale, and the scarcity became more grievous and excessive than ever" (*Death of the Persecutors* 7). Of course, a black market, with inflated prices, appeared under the noses of the authorities. The environmental base of the Roman world continued to deteriorate, and the system Diocletian envisioned collapsed along with the unity of the empire.

Conclusion

It must be remembered that the Roman economy was based on the agrarian sector, and that agricultural productivity was a crucial factor. The inevitable result of the human failure to support nature was that nature could support fewer human beings. Population decline was a continuing problem in the later imperial period. Emperors time and again tried to counteract it by making marriage and childbearing mandatory for citizens. Declining population meant fewer farmworkers, so that reductions in population and agricultural production tended to be synergistic. Diocletian's edict on occupations, requiring men to provide sons to fill their positions, indicates that there was a shortage of labor. Although periodic plagues were also to blame for the decline in population, the chronic agricultural decline was the basic problem, and it derived from environmental causes.

Social organization and technology can be used either for positive or negative purposes ecologically. The Romans used them in both ways, but unfortunately the trend over the centuries was destructive. Nonrenewable resources were consumed, and renewable resources were exploited faster than was sustainable. As a result, the Mediterranean lands were gradually drained, losing a large portion of their living and nonliving heritage. This was the fate of the natural environment and human populations alike, and it was not something that came irresistibly from outside with a climatic change or other natural disaster; it was the result of the imprudent actions of Romans themselves.

The conclusion that must be drawn is that the structure of the society and economy of the Romans caused environmental changes that depleted their natural resources and were of critical importance in hampering their ability to feed the population, to maintain health, and to prosper. These changes therefore weakened society, depleting its human resources. Their effects were felt early, but were cumulative, reaching a devastating level that plagued the empire in the centuries that led to its fragmentation.

CASE STUDY C: THE ASWAN DAMS AND THEIR ENVIRONMENTAL RESULTS

When the King says it is midnight at noon,
The wise man says, "Behold the moon."

—OMAR KHAYYAM, *RUBAYYAT*,

TWELFTH CENTURY AD

One of the twentieth century's greatest technological projects, and one with far-reaching environmental effects in the Mediterranean basin, was the construction of the Aswan High Dam on the Nile River. Officially completed in 1971, it has been acclaimed as Egypt's greatest modern national asset and criticized as an ecological disaster, and has aroused more controversy than any other resource development project in the region. It illustrates the important but often forgotten principle that large dams do not just provide electricity, irrigation, and flood control but also have a host of environmental effects, some of them unintended and unfortunately many of them negative. Assessment of these effects shows mixed results and indicates a missing element in many large development projects: a careful examination of perspectives that could be provided by environmental history.

Traditional Agriculture

Aswan was a key to the Nile in ancient as well as modern Egypt. It is the location of the First Cataract, a rapid in the river that presented the first block to navigation as boats proceeded upstream. Here granite rocks, representing a very early period in the geological history of the Earth, pierce the water's surface, resisting the erosive power of the river. Nearby, the Egyptians opened a granite quarry from which great monolithic obelisks were taken to adorn temples at Karnak and Heliopolis. Egypt depended on the annual flood of the Nile to water the soil and produce the crops on which the people lived. A device to measure the depth of the flood, called a nilometer, was constructed at Aswan so that those downstream could be warned of the depth of the coming inundation.

The Aswan High Dam across the Nile River, which generates electrical power and has created Lake Nasser, which holds a volume of water equal to more than two years of the Nile's average flow. Aswan, Egypt. (Carl and Ann Purcell/Corbis)

The ancient Egyptians recognized that the gods had divided their part of the Earth into contrasting terrains: the Red Land of the desert and the Black Land along the Nile. There are two major ecosystems in Egypt, desert and riparian, and the Egyptians must live primarily within one of them, the riverine. The only arable land was what the river reached in the narrow valley of Upper Egypt and the wide delta of Lower Egypt, which together make up less than 5 percent of the territory; except for a few oases, the rest is rainless desert. For thousands of years, the annual Nile flood watered Egypt's cropland and left behind a layer of soil-building silt. Ancient Egypt is a model showing how a people could live within a limited ecosystem and prosper without destroying the cycles that supported it and them. Sustainable agriculture cooperated with the river's rhythm. Human labor directed the flood, whenever it rose high enough, into a series of basins where its depth and the length of time it remained on the fields could be controlled to a great extent. As William Willcocks, father of the first Aswan Dam, put it, "The ancient Egyptians understood thoroughly how to utilise the flood waters of the Nile" (Willcocks 1903, 25). The result of the time-honored balance of Egyptian agriculture and the environment was a supply of food and fiber that was sufficient, at least in most years. The river, although regular, was not totally predictable; there were years of a low Nile when crops were short, and years of a high Nile that swept away villages and irrigation works. No wonder the ancient Egyptians prayed to the river for the blessings of an ample but moderate flood. In most years they got it, and there was food to store for lean years and to export without depriving the Egyptians. That order of things persisted down to the nineteenth century, when one could still see a landscape and witness human activities little changed since pharaonic times.

The First Aswan Dam

The transformation of Egypt from a society dependent on traditional agriculture to an adjunct of the world market economy began with Muhammad Ali, who ruled Egypt in the early nineteenth century. Desiring to enrich himself through commerce, he set up factories and in 1822 began the cultivation of cotton in the delta. Cotton was primarily for export, and it could be grown as a second crop on land already in cultivation, but it required irrigation during the time of the year when the Nile was lowest. The problem could be solved by building barrages, that is, low dams that raised the level of the river enough so that part of the flow could be diverted into canals. Muhammad Ali laid the first stone for a double barrage at the point where the Rosetta and Damietta branches of the Nile diverge north of Cairo in 1833. He was barely dissuaded from using stones from the great pyramids to build it (Hurst 1957, 50), and the construction using quarried stone was carried on between 1843 and 1861. Ismail Pasha, a successor of Muhammad Ali, profited from the new agricultural policy when the U.S. Civil War deprived world markets of their former largest source of cotton for several years. Egyptian cotton exports rose from 17,500 tons in 1850 to 150,000 tons in 1880, an increase of 750 percent. The effect on Egyptian agriculture was immense. The cultivated area, 3.05 million feddans in 1813, had expanded in 1877 to 4.74 million (in Egypt, the unit of land measurement is the feddan, equivalent to 1.038 acre or 0.42 hectare). Most important, the transition from basin to perennial irrigation began in earnest. It would transform the landscape and the ecology of the soil, and create a demand for "timely water," that is, water for cash crops in the former fallow season.

The British, who seized control in 1882, were more than willing to forward for their own benefit the agricultural revolution begun by Muhammad Ali. The government of Sir Evelyn Baring, Lord Cromer, brought in a highly skilled staff of water engineers and put them to work repairing the irrigation infrastructure, which had badly deteriorated. They strengthened the barrages, dredged out the canals, and tried to disentangle the lines of irrigation from those of drainage. They knew drainage was important, since perennial irrigation without it tended to waterlog the soil, to raise the water table, and to cause salinization. It should be noted that in the last years of the century, during the months of the low Nile, virtually no water flowed into the Mediterranean Sea at the river mouths in the delta.

In 1894, Sir William Willcocks, born and educated in India, proposed an "open dam" at Aswan that would create a reservoir holding 2.4 billion cubic meters of water. This amount would be available for irrigation during the cotton-growing season. Willcocks was aware of the value of the Nile sediment to the cultivated fields. As he later remarked, "It will be an evil day for Egypt if she forgets that . . . the lessons which basin irrigation has taught for 7,000 years cannot be unlearned with impunity. The rich muddy water of the Nile flood has been the mainstay of Egypt . . . and it can no more be dispensed with today than it could in the past" (Waterbury 1979, 39). He knew that his relatively small reservoir would fill with silt in very few years unless engineers designed the dam to let through the first part of the annual flood, which contained almost all the silt. He had experience with the Periyar Dam in South India, designed with a sluiceway near its base to let the muddy floodwaters through. The last portion of the flood could be impounded because the water then was relatively clear and would not deposit detritus in the reservoir. So his plan for Aswan included sluices with cleverly designed gates that could be opened to let the muddy water through. There were to be 140 of these at the bottom of the dam, and 40 more at a higher level.

The project was delayed for four years while Lord Cromer looked for funding. Sir Ernest Cassel, multimillionaire financier and friend of King Edward VII, offered a loan. He was not disinterested, since his company had Egyptian land investments. In the meantime, another issue agitated the world of arts and letters: the dam would periodically flood the temples on the river island of Philae. Built by the Ptolemies to honor the goddess Isis, these were among the most graceful and complete buildings surviving in Egypt from antiquity. At a time when money for the dam itself was at a premium, the cost of moving the temples out of the way was thought prohibitive. Winston Churchill, then a young officer on an expedition to the Sudan, thought the Philae temples should be sacrificed "to the welfare of the world" (Addison 1959, 43). They were eventually moved at the time the Aswan High Dam was constructed in the 1960s, as part of the UNESCO project to "save" the monuments. Little concern was voiced for the Nubian villagers whose lands and homes would also be flooded. Sir Benjamin Baker, the consulting engineer, scaled down Willcocks's dam by 20 feet, supposedly to protect Philae, although water in fact entered the temples during high water. This lowering reduced the maximum volume of the reservoir to one billion cubic meters. Construction began in 1898 and was completed a year ahead of schedule, in 1902. The dam was 1,940 meters (6,400 feet) long and nearly 20 meters (65.5 feet) high, its total volume somewhat less than half that of the Great Pyramid. Four locks on the west side of the dam permitted river traffic to bypass it.

This first Aswan Dam reserved the end of the flood and spread it out over a longer time, but reduced its height. Barrages in addition to those begun in the Muhammad Ali period raised the river and made its water available at a higher level. The Asyut barrage, 345 miles below Aswan, was built at the same time as the dam. It consisted of a wide masonry platform carrying a bridge of 111 arches, each 5 meters (16.5 feet) wide and fitted with two gates. New barrages arose at Zifta on the Damietta Branch (1903); at Esna, one hundred miles below Aswan (1909); and at Nag Hammadi below Luxor (1930).

Soon the managers discovered that the low dam did not retain enough water to supply the business projects of its backers. Cassel's company had purchased some desert land at Kom Ombo north of Aswan, and the water supply from the dam proved insufficient to irrigate it. The archaeologists' opposition could not hold back the compelling arguments of commerce, so between 1907 and 1912 the dam was raised about 7 meters (23 feet), to 27 meters (88.5 feet), to increase its holding capacity approximately to the size Willcocks originally had recommended (Mansfield 1971, 118).

A new commission in 1920 recommended that two dams be built in the Sudan to supplement flood control at the Aswan Dam and to provide water for the Sudan's Gezira Cotton Scheme. The Sennar Dam on the Blue Nile was completed in 1926, followed by the Jebel Auliya Reservoir south of Khartoum in 1937. (The latter, on the White Nile, is a white elephant, as water engineers realized when measurements showed that while it stores 3.6 billion cubic meters, it evaporates 2.8 billion cubic meters per year.)

The demand for irrigation in Egypt continued to grow, and since it had become evident that Nile water would have to be shared with Sudan, irrigation experts proposed a second heightening of the Aswan Dam to store more of Egypt's water in Egypt each year. Between 1929 and 1933 the Aswan Dam was raised yet again, this time by 9 meters (29.5 feet), to 36 meters (approximately 118 feet) (Smith 1971, 221). The reservoir capacity more than doubled, to something over five billion cubic meters, or 6 percent of the average annual discharge.

The earlier Aswan Dam, with its two heightenings, produced many, but not all, of the side effects that later were to appear with the High Dam. As noted above, the engineers chose a design that would allow the annual flood, with its deposition of silt, to continue. Even so, there was some downstream scouring and lowering of the riverbed.

The dam flooded part of Lower Nubia every year, displacing an increasing number of Nubians with each raising of the structure. These people relocated themselves to other villages north of Aswan, or to Cairo. In large part settlers engaged in nonagricultural pursuits such as trade and transport; many of them were boat-owners who had clustered close to Aswan to conduct their business. They received small compensation, but no aid in relocation.

The worst environmental problems, such as inadequate drainage, waterlogging, and salinity, had appeared as early as 1890 and worsened after 1902. The need to address the long-standing drainage problem was clear. In some areas, the rate of schistosomiasis, a disease transmitted through water with snails as an alternate host, rose from 21 percent to 75 percent (Lanoix 1958, 1011–1135). These were results of the shift to perennial irrigation accelerated by the dam. Silt deprivation and losses of nutrients also were noticed, though not to the extent that would occur with the High Dam.

Retreat of the coastline at the outlets of the Nile and invasion of the delta by seawater were noted after the 1933 heightening of the Old Aswan Dam. Works to counter coastal erosion were begun, but over the years they proved ineffective. What ultimate good would a dam be if Egypt were to lose a major part of the Nile Delta?

The Decision to Build the High Dam

The Egyptian revolutionary government of Gamal Abdel Nasser announced in 1952 the decision to build a high dam on the Nile River at Aswan. Interestingly enough, Nasser made the announcement in the year of the fiftieth anniversary of the completion of the first modern dam at Aswan. The Aswan High Dam represents a massive break with the past, and as in all such cases, it had results beyond those intended by its planners and builders. Eclipsing the earlier dam built by the British in 1902 and raised in 1912 and 1933, it was a gigantic step in the ongoing transformation of Egypt from a society dependent on traditional agriculture to an adjunct of the world market economy. It rises 110 meters (364 feet) above its base at the head of the First Cataract, its length across the river valley is more than 2.33 miles, and its reservoir, Lake Nasser (called Lake Nubia in Sudan), can hold two years' average flow of the Nile. The dam ended the annual flood and converted the river below it into an open aqueduct.

There were several purposes of Nasser's decision to impose such a traumatic alteration on the Nile and the people who depend on it. Egypt is unique in that all its cropland is irrigated. The dam, its proponents believed, would make perennial irrigation, and a second or even third crop, possible on all Egypt's cultivated land. The additional production would be in export crops, convertible into cash on the world market, especially cotton, sugar, and rice, and also maize and wheat. In addition, there would be extra water to expand cropland by as much as two million acres. Maintaining food production to feed Egypt's growing population was undoubtedly one of the intended goals of this expansion, but a secondary one can be gathered from the fact that Egypt, largely self-sufficient in food production before the building of the High Dam, imported 70 percent of its food at the end of the twentieth century. Exports compensated in part for these massive imports.

A second function of the dam would be generation of electric power for industrialization. Optimistic predictions put the generating capacity at ten billion kilowatt-hours, and substantial growth in manufacturing was expected. After the completion of the High Dam, it generated about 50 percent of all the power used in Egypt. By the 1990s, that figure had declined to 10 percent because other power stations had come online.

A third purpose was flood control; the vast size of the planned reservoir allowed for containment of even large floods, and flooding on the Nile below Aswan was to be ended. Of course, any astute hydraulic engineer would have known that these three purposes would more or less interfere with each other. It is impossible to maximize two independent variables over time, much less three. Irrigation would require releases of water at times that are not optimum for power generation, and vice versa. Power generation is most efficient with a full reservoir, and flood control requires a lower level with room to receive surges from upstream, and so on.

There were also, however, persuasive political purposes for building the dam. John Waterbury wrote that "Egypt is the epitome of the downstream state" (Waterbury 1979, 5). The Nile, the world's longest river, flows 6,611 kilometers (4,132 miles) from its southernmost source to the Mediterranean Sea. Only the last 1,520 kilometers (950 miles) are within Egypt. A former Nile development plan backed by H. E. Hurst, a director general in the Egyptian Ministry of Public Works, and adopted by the government in 1948 was termed the "Century Storage Scheme" because it would allow for the extremes of high and low water that could be expected in a hundred-year period, and would have kept the Old Aswan Dam and added a series of dams, reservoirs, and canals in the upstream states. At the time of the decision, the headwaters of the Nile were in British hands. The later division of the upper watershed into nine independent nations, several chronically unstable or potentially hostile, has offered Egypt little improvement in prospects for a basin-wide agreement to guarantee the annual flow that is its lifeline. The upstream states that supply water to the Nile include, for the White Nile, Burundi, Rwanda, Zaire, Tanzania, Kenya, and Uganda; and for the Blue Nile, Ethiopia, which is also the source of two other main tributaries, the Sobat-Baro and Atbara. (A minor fraction of the Atbara flows from Eritrea.) It is noteworthy that Ethiopia provides 86 percent of the water that reaches Aswan. The High Dam allows Egypt to control storage, granted that part of the reservoir is in the Sudan and makes cooperation between the two states essential.

Another political purpose of the High Aswan Dam was to be fulfilled by the grandeur of the project. The dam as built is seventeen times the volume of the Great Pyramid (Goldsmith and Hildyard 1984, 1). It would be a lasting monument to Nasser, to the revolution, and to Egypt's independence.

It was no wonder that Nasser quickly announced in the year he seized power that it would be built. Within four years, when the West reneged on promises of technical and financial aid, Nasser seized the Suez Canal with the intent of using its revenues for the dam, fought a war, and turned to the Soviet Union for assistance. In that context, discussions of possible negative effects had to be conducted with circumspection. One technician who was there at the time later expressed the situation by quoting the *Rubayyat* of Omar Khayyam: "When the King says it is midnight at noon, the wise man says, 'Behold the moon'" (Waterbury 1979, 101). Egyptian and Western engineers did ask a number of questions concerning possible effects of the dam on stream flow, removal of the silt load and erosion of channels downstream, evaporation from the reservoir, sedimentation, degradation of the delta coastline, induced seismic activity, and seepage during the early days of planning. But since the government was committed to the dam, it became less receptive to these questions even though answers to them might have affected the project design, and discussion was "discouraged or suppressed" (White 1988, 8). Dr. Abd al-Aziz Ahmad, chairman of the Hydroelectric Power Commission, criticized the project on the grounds of excessive water loss through evaporation, presenting his conclusions in British journals in 1955–1960. Egyptian leaders thought he was offering ammunition to adversaries abroad, and his opinions, no matter how well considered, subsequently met deaf ears. He suffered disgrace and professional ostracism. Ali Fathy, supervisor of the Old Aswan Dam reservoir and professor of irrigation at Alexandria University, remarked, "It became clear that competent technicians in government circles were collectively determined to overlook any signs of the deterioration of soil fertility as a side effect of the High Dam, even as a hypothesis. This was the result of what might be called the 'High Dam Covenant,' a psychological state born of political and other circumstances which . . . cloaked the project from its very inception" (Fathy 1976, 50–51). If those who raised technical questions risked their careers, those who warned of negative social or environmental consequences had even more to fear. As Hussein Fahim put it,

> Government policy was not to be debated publicly before being formally adopted. Policies were to proceed from the top downward. The open public channels of technical and political dialogue were blocked. Actually, anything that was described less than superlatively became potentially treasonous. As a result, the reasonably balanced combination of the political and the technical in the execution of big development schemes, designed to avoid the waste of scarce resources, was undermined. . . . [T]he late 1950s and the entire following decade witnessed a total blackout of any [discussion of] mistakes or malfeasance connected with the Aswan High Dam. (Fahim 1981, 165)

Foreign consultants also muted criticisms. An International Bank for Reconstruction and Development "review of its own involvement in the scheme revealed that the ecological ramifications of the dam . . . did not figure prominently in its own positive evaluation of the project" (Waterbury 1979, 102).

Could the planners who considered building the High Dam have avoided some of the worst mistakes in this situation of running up against inexorable limits? The modern environmental history of Egypt, including the first Aswan Dam and its heightenings, could have provided warnings that might have helped prevent some of the damaging effects of the High Dam, or led to a decision not to build it.

The Consequences of the Decision

Gilbert White termed the construction of the Aswan High Dam "a massive, unique intervention in physical, biological, and human systems" (White 1988, 38). It was not unique in that other huge dam projects occurred before and after it, even in Egypt. It had precedents in environmental history. But it is true that it was a massive intervention, and that such an intervention will always have unintended consequences, some of them foreseen and some not. Some of the important side effects of the Aswan High Dam should be noted.

Although management of the reservoir envisioned a flood control capacity of forty-seven billion cubic meters, it became clear that if an unusually high flood came down the Nile while the reservoir was near capacity, facilities at the dam might not have the capacity to handle it without a disaster. This danger was anticipated by the construction in 1981 of the Toshka spillway, 248 kilometers (155 miles) south of the dam, which can send water into the Toshka depression (Amin 1994). The diverted volume would be subject to evaporation, and unavailable for irrigation.

A predicted effect of the dam was the loss of a fraction of the water stored in the reservoir in a hot, dry, windy environment. Estimates by various authorities of the amount lost annually vary between nine and seventeen billion cubic meters, with the best guesses clustering around twelve to fourteen billion, that is, about 15 percent of the average annual flow of the Nile at Aswan. Another effect of evaporation is to increase the salinity of the stored water; the salt content of the river entering Lake Nasser is 200 parts per million, but that of the water leaving it is 220 parts per million.

Experts knew that a large amount of water would seep into permeable rock formations surrounding the lake. They hoped that the annual amount lost, though huge at first, would decline due to saturation of the strata and sealing of the reservoir by silt deposition. The saturation occurred, at least partially, but there was no sealing by silt, since all the sediment is deposited at the upper end of the lake. Estimates of the amount presently lost differ widely, but five billion cubic meters is probably conservative. Combined with evaporation, this would mean a loss of 20 percent of the river's annual flow.

Some planners worried that the weight of filling Lake Nasser would place strains on faults in the rock formations and that seepage would also lubricate the faults, in both ways causing earth tremors. An earthquake registering 5.5 on the Richter scale, possibly induced by the reservoir, did occur on November 14, 1981, without major damage to the dam (Abu-Zeid et al. 1995). Smaller quakes have occurred, and there has been minor shifting of the body of the dam, but experts estimate that an earthquake registering seven on the Richter scale would not threaten the integrity of the dam, and that the probability of such an event is extremely small.

The most far-reaching ecological effect of the dam is that the entire load of sediment and nutrients carried by the river settles out in the reservoir. This occurs where the river enters the lake; it has begun to form a new delta there at the rate of 130 million tons of material each year, and will in time fill the lake. An often-repeated estimate of the useful life of the reservoir is five hundred years. Oddly enough, Sudan may be first to find a bonanza of rich soil at the head of Lake Nubia. An unpredicted process at the reservoir was the deposition of windblown sand and dust. Sand dunes blow out of the desert and spill into the lake at a rate equal to one-tenth of the accumulation of river-borne silt. This occurs all along the western shore and alters the chemical composition of dissolved solids in the lake and the river downstream from the dam (El-Moatassem 1994; Stanley and Wingerath 1996). The water emerging from the dam is virtually free of suspended solids.

Many engineers pointed out that the erosive power of flowing water is greater if it is not loaded with silt; thus the river below the dam would scour and lower its bed, affecting the stability of the barrages and bridges in Upper Egypt and making reinforcement necessary. These predictions proved true. Lowering of the bed of the Nile increases the difficulty of getting the water from the river into canals. In addition, erosion decreased water depth at entrances to locks and near bridges, causing navigation problems and requiring dredging (Gasser and El-Gamal 1994, 36). Caving, cracking, and sliding of banks has been observed as far as one hundred kilometers (62 miles) downstream from Aswan, and this is particularly serious because cultivation generally extends to the river's edge.

The Nile Delta is the result of silt deposition. This is what Herodotus meant when he wrote, "Egypt is a gift of the Nile" (*The Persian Wars* 2.5). With virtual elimination of the sediment load from the river where it enters the Mediterranean Sea and is used to replenish the delta, coastal erosion might well have been expected, and it was, although few steps were taken to counter it before the High Dam was built. Now the delta shoreline at the river mouths is retreating at an average yearly rate of thirty meters (ninety-nine feet). Since 80 percent of Egypt's agricultural land is in the delta, along with 75 percent of the nation's population, loss of land there is a serious threat. The coastline is not stable; geologically, the delta is sinking under its own weight, at a rate between three and fifty millimeters per year (Theroux 1997, 8). The sea level is rising: since Roman times, the Mediterranean has risen two meters (6.6 feet). This brings the additional danger of invasion by seawater; in the north Nile Delta, underground water is brackish, making wells useless in some places. The bars protecting coastal lakes are eroding and could collapse, allowing further invasion by the sea. The delta is the only area of Egypt with extensive areas of wetland inhabited by birds and other wildlife, and both the encroachment of the Mediterranean and works constructed to slow it are likely to have an impact on them.

Before the High Dam was constructed, nutrients brought to the Mediterranean by the Nile supported a plentiful population of fish. In the early 1960s, the sea fishery brought in an annual catch of twenty-five thousand tons, half of that consisting of sardines. After the dam was finished, the catch of sardines declined by 88 percent, and that of bottom fish by 64 percent, even though the number of motorized boats doubled and the fleet traveled much farther in search of fish. The even more productive fishery on Lake Burullus in the north delta fell from sixty thousand tons in 1966 to fourteen thousand in 1975. In the Nile itself, besides a decline in number of fish taken, their size was smaller on average, and the number of species fell; at Mansoura in the delta, for example, from eleven species to three, and at Asyut in Upper Egypt from forty-seven to twenty-five. There was a partial compensation for Egypt's fishing industry, if not for the riverine ecosystem, in that the take from Lake Nasser had increased from under eight hundred tons to twenty-two thousand tons by 1978.

An unforeseen effect of the loss of sediment was on building. Formerly, bricks were made from the annual deposit of silt. Now that was no longer possible, and brick makers turned to the delta, paying peasants what for them was a handsome sum to strip the topsoil from their land to be made into bricks. In the 1980s, brick-making consumed 120 square kilometers (48 square miles) of topsoil annually. Without a source of renewal, this amounted to liquidating Egypt's agricultural capital. It has been illegal since 1984, but continues at a smaller rate in spite of that.

The shift from basin irrigation to perennial irrigation was one of the intended purposes of the High Dam and was greatly accelerated by it. Almost a million acres were converted by 1975. Basin irrigation was used throughout Egypt before the nineteenth century. This system divided the land into basins of from one thousand to forty thousand feddans by means of banks. When the Nile rose, water was allowed into these basins successively, up to one or two meters in depth, and held there until the flood had fallen enough to release it back to the river, about forty to sixty days. During that time it dropped its silt, forming a flat surface for cultivation.

Perennial irrigation, virtually universal in Egypt today, runs water through a network of canals onto the land at regular intervals of two or three weeks throughout the year. It requires a year-round supply of water, and barrages to raise the Nile so that water can be conducted into the canals. Egyptian farmers are often criticized for overwatering, but perennial irrigation almost inevitably creates problems of waterlogging unless adequate drainage is provided. Unfortunately, a 1958 project to install main and field drains was abandoned because planners thought that with a lower Nile and no flood, drainage would improve enough to make them unnecessary. The opposite happened. The water table has risen from a former average of fifteen meters (49.5 feet) below the surface to three meters (9.9 feet). In Cairo, the water table is only 81 centimeters (32 inches) below the surface (Fahim 1981, 31). Ninety percent of the cultivated land in Egypt is waterlogged. The increased evaporation has contributed to a rise in average relative humidity, which along with soil moisture is a danger to the ancient monuments.

Waterlogging is related to another extreme problem, salinization. When water is allowed to evaporate from land without adequate drainage, the salts it contains will accumulate in the soil. In many places in Egypt, salt is visible on the surface as a white crust, and the rate of accumulation is as much as a ton per hectare per year (Pearce 1994, 30). The United Nations Food and Agriculture Organization has reported that 35 percent of Egypt's cultivated surface is salinized. A 1982 study estimated that 10 percent of Egypt's crops are lost annually to deterioration of soil fertility (White 1988, 36). It was noted above that evaporation in Lake Nasser increases salinity in the Nile marginally; unfortunately, improved

Agriculture in the Fayum Oasis, below sea level, is irrigated by water from the Nile. Evaporation has left crystals of salt on the soil in the foreground, an example of salinization that makes it unusable for crops. (Photo courtesy of J. Donald Hughes)

drainage would also increase it. Salt in the Nile at Cairo before the High Dam was constructed was measured at 170 parts per million; today it is 300 parts per million. More salt is now reaching the delta, and less is reaching the sea. In the western delta, where the water table is rising to the roots of plants, salt concentration reaches 3,000 parts per million. (For comparison, by treaty the United States cannot deliver water containing over 1,000 parts per million to Mexico; in the late twentieth century the level in the lower Colorado River was 1,500 parts per million.) Recently it was decided to install tile drainage on over two million feddans, about one-third of Egypt's cultivated land. If this vast project is undertaken, it will take land out of use during construction.

Since perennial irrigation provides no silt and nutrients to the land, nitrogenous and phosphate fertilizers are applied at a rate that has increased exponentially. To reduce dependency on imports, a factory was built at Aswan; it uses the entire output of the electrified Old Aswan Dam, or two billion kilowatt-hours. Drainage water is polluted by undesirable levels of fertilizers and pesticides, and yet is the source of additional supplies that Egypt hopes to pump to new lands, such as those being developed in Sinai.

An unexpected result of fertilizers in the water was eutrophication. Phytoplankton, stimulated also by the sunlight that easily penetrates the now clear water, and not flushed out to sea by the flood as in the past, has reproduced to an enormous degree and has clogged the water purification system in Cairo. Hydrophytes such as water hyacinth, a plant introduced from South America, have opportunistically covered the surfaces of 82 percent of the watercourses, canals, and drains. Mahmoud Abu-Zaid calls this unforeseen "sudden weed invasion . . . perhaps the most serious side effect of the construction of the High Dam" (Abu-Zaid and Saad 1993, 37). He estimated that the transpiration caused by hyacinths increases the annual evaporation by 7.4 billion cubic meters. They have been fought by mechanical clearing and by massive applications of herbicides in Egypt, which have in turn destroyed many nontarget plants and animals.

A much-discussed possible effect of the dam was an increased incidence of schistosomiasis (bilharzia), a disease caused by parasitic trematode worms that pass into water in urine and feces and infect certain snails as alternate hosts. The disease is very debilitating and can lead to death. It is common in warm countries where people work in contact with water for long periods of time, which is certainly true of Egypt. It has been present in Egypt at least since pharaonic times; the eggs of the worm have been found in the kidneys of mummies from as early as the Twentieth Dynasty (1186–1069 BC), and the disease was probably present long before that. There are two forms of the parasite in Egypt. Urinary schistosomiasis, previously the common form, has decreased in Egypt as a whole since the building of the dam due to public health measures such as protected water supplies and rural health centers, although it has increased in some localities. The more severe form, intestinal schistosomiasis, however, has spread in the delta, and has appeared in Upper Egypt. The reason for the increased danger of contracting schistosomiasis is not the dam itself, but the conversion to perennial irrigation. Under basin irrigation, drying during the season of low water controlled snail numbers, but now snails have a permanent habitat. Also, perennial irrigation keeps people in contact with the water throughout the year. One study said that between Cairo and Aswan, the urinary form increased from 5 percent in 1930 to 35 percent in 1972. A University of Michigan study found declines in that form in all major districts due mainly to the use of protected water supplies. But the more severe variety, intestinal schistosomiasis, spread and increased from 3.2 percent in 1935 to 73 percent in 1979 (Fahim 1981, 34). Malaria has become more of a problem in Nubia and the Sudan with the increase in slack water.

It was clear from the moment that the Aswan High Dam was approved that there would be immense costs to the people and land of Nubia, which was almost completely inundated by the reservoir. As a result, more than 110,000 Egyptian and Sudanese Nubians had to be resettled. Water drowned their towns and villages, and the land itself, with its productive fields and date palms, disappeared under the lake. The Nubians, a people adapted to the riverine ecosystem of the Nile between the First and Third Cataracts through thousands of years of continuous occupancy, would have to endure a sacrifice on behalf of the prosperity of the much more numerous people downstream. The Egyptian government provided resettlement, with education, health care, and land, in "New Nubia," Nuba al-Gedida, north of Aswan and east of Kom Ombo. For the Nubians, it was a foreign ecological environment, a flat landscape with no palm groves, and worst of all, too far from the Nile. Housing was of an unfamiliar style. The terms of their resettlement required them to raise sugar cane, a crop they knew little about, by a rigorously timed water distribution. Sudanese Nubians were moved to new settlements near Khashm el Girba, a dam built in 1964 on the Atbara, a tributary of the Nile that flows from Ethiopia (Collins 1990, 272). They were given community services and leased land, and directed to raise cotton, wheat, and peanuts in a three-year rotation scheme. In addition, the land was already in use by animal-raising nomads who resented the intrusion into their traditional territory, and government attempts to sedentarize the latter and to settle them near the new residents met with only partial success.

Many Nubians chose to go elsewhere than the resettlement areas. Some sought jobs in Cairo and Khartoum. Others refused to leave Nubia, or returned there after a time in the settlements. They developed agriculture near the reservoir with subsoil water and lift irrigation. Some provided tourist services, chiefly at Abu Simbel. They did not take up fishing because fishermen from the Aswan area had already moved their boats onto the lake and established a monopoly there. In Sudan, many of the Wadi Halfa people moved to a poorly planned new town at the railroad terminus on the west shore of the lake.

Too well-known to be retold here is the story of the flooding of thousands of archaeological sites in the richly historical land of Nubia, many of them large, well-known, and hitherto well preserved. A vast international effort coordinated by UNESCO moved the major monuments, including the colossal temples of Abu Simbel, and salvaged other sites, but an untold number were lost, perhaps forever.

One of the announced purposes of the High Dam was to open new areas for cultivation to compensate for the flooded fields in Nubia and to increase production. The desire for reclamation is understandable because Egypt has a population density of more than a thousand people for every square kilometer of cultivated land. When the government approved the construction of the High Dam, there were 6 million feddans under cultivation, and the target figure was 1.3 million new feddans to be brought into production. In 1977, the goal was raised to roughly 2.8 million feddans. By 1982, work had started on approximately 1 million feddans, but only half of that was actually farmed and less than 20 percent was irrigated. The total acreage of irrigated land had declined due to urban expansion, brick-making, waterlogging, and salinization. Productivity increased, however, due to the shift of 912,000 feddans from basin to perennial irrigation.

Government predictions for newly reclaimed land are excessively optimistic because they often assume fertility equal to old lands, thus failing to take into account the fact that soil is not just a physical substrate in need of water, fertilizer, and seeds, but a living ecological community. Heavy desert soils will not produce without great expense of energy, materials, and time. The quality of soils in new lands is generally poor; 70 percent are Class IV soils. Many of the valuable export crops, such as cotton, wheat, rice, and sugar cane, are not successful on the new lands, although other crops often are, for example citrus, alfalfa, potatoes, and sugar beets. Still, Egypt goes on with projects to divert water to the Sinai and the desert depressions without a clear sense that water may, one day, prove inadequate in quantity and/or quality.

The environmental effects of the Aswan High Dam were not entirely hypothetical at the time of the decision to build it. But in authorizing the High Dam, the historic negative effects of the older dam were not given serious consideration, since those who could have commissioned studies were already committed to the project. Historical precedents were available not only for dams in other environments but also for dams on almost the same site on the Nile itself. As John Waterbury observed, "The history of this project is testimony to the primacy of political considerations determining virtually all technological choices with the predictable result that a host of unanticipated technological and ecological crises have emerged that now entail more political decisions" (Waterbury 1979, 5–6). He termed Egypt's policies leading up to the dam "shortsighted" and "non-integrated."

Conclusion

The antidote for shortsightedness is careful consideration of both environmental history and the need for sustainability in the future. The antidote for a non-integrated approach is consideration of the many facets of the ecosystem, including the fact that humans cannot control every aspect of it, since massive actions always have massive unintended effects, nor can humans exceed the limits of the ecosystem without catastrophic results for themselves.

At least two problems lessen the possibility that Egypt can arrive at a sustainable level of production within the limits set by water, land, and the Nile Valley ecosystem. The first is population. At the time the Old Aswan Dam was under construction, Egypt had ten million people. With the High Dam rising, the population passed thirty million. In 1995 it was sixty-three million, heading toward ninety-seven million in 2025, in spite of one of the lowest growth rates in Africa. This pattern indicates an expanding demand for water in the future. Where will it come from?

Second is urbanization. Every year a larger percentage of Egyptians live in cities, particularly Cairo, which had 7.5 million people in 1976 and passed 17.3 million in 2000, containing 25 percent of Egypt's population. In the same period, Alexandria grew from 2.5 million to 6.6 million. Industrial, commercial, and residential building, with urban infrastructure, will use increasing amounts of space and water, in spite of a 1984 law prohibiting urban development on agricultural land. Estimates indicate a water deficit for Egypt of fourteen billion cubic meters by 2025.

The Nile will not grow in the future, but upstream projects might send more water to Egypt. Most ambitious is the partially constructed Jonglei Canal in southern Sudan, intended to carry water past the Sudd swamps and to end the evaporative loss of half the flow of the White Nile, but now halted by war since 1984. By drying up a huge wetland, Jonglei would damage a unique ecosystem and decimate wildlife. Sudan has treaty rights to half the additional Jonglei water and will undoubtedly use it if it becomes available. What of the other upstream states? Ethiopia's population, growing at twice Egypt's rate, will soon surpass Egypt, and could reach 127 million in 2025. William Willcocks once remarked, "it might not be convenient on political grounds to put one of the great public works of Egypt at the absolute mercy of the Abyssinian Emperor" (Willcocks 1903, 9). At different times subsequently, when Ethiopia proposed irrigation projects using the headwaters, Prime Ministers Nasser, Anwar Sadat, and Hosni Mubarrak each threatened war with any state that takes "Egypt's water" (Fahim 1981, 160). Sooner or later an international plan for the watershed must be negotiated, possibly through the United Nations. But no plan can meet the desires of every nation concerned to support its growing population and to achieve economic growth by producing more for world markets.

As far as the decision to build the Aswan High Dam is concerned, if the experience of the past had lessons to teach, they seem not to have been learned. The present ecological situation of Egypt is precarious. It is difficult to imagine what the path to sustainability might be, since the constraints of politics convince planners, and they almost never consider the limits of the ecosystem. But planning will be misleading until it takes account of perspectives provided by environmental history.

References

Abu-Zaid, Mahmoud, and M. B. A. Saad. 1993. "The Aswan High Dam, 25 Years On." *UNESCO Courier* (May 1): 37.

Abu-Zeid, M., W. A. Charlie, D. K. Sunada, and A. Khafagy. 1995. "Seismicity Induced by Reservoirs: Aswan High Dam." *International Journal of Water Resources Development* 11, 2 (June): 205–213.

Addison, Herbert. 1959. *Sun and Shadow at Aswan: A Commentary on Dams and Reservoirs on the Nile at Aswan, Yesterday, Today, and Perhaps Tomorrow.* London: Chapman and Hall.

Amin, K. 1994. "Safety Considerations, Operation, and Maintenance of Existing Structures of High Aswan Dam." *International Water Power and Dam Construction* 46 (January): 40–41.

Aubert, Jean-Jacques. 2001. "The Fourth Factor: Managing Non-Agricultural Production in the Roman World." In *Economies beyond Agriculture in the Classical World,* eds. David J. Mattingly and John Salmon, 90–112. London and New York: Routledge.

Charlesworth, Martin P. 1951. "Roman Trade with India: A Resurvey." In *Studies in Roman Economic and Social History,* ed. P. R. Coleman-Norton, 131–143. Princeton: Princeton University Press.

Collins, Robert O. 1990. *The Waters of the Nile: Hydropolitics and the Jonglei Canal, 1900–1988.* Oxford: Clarendon Press.

Duncan-Jones, Richard P. 1990. *Structure and Scale in the Roman Economy.* Cambridge: Cambridge University Press.

El-Moatassem, M. 1994. "Field Studies and Analysis of the High Aswan Dam Reservoir." *International Water Power and Dam Construction* 46, 1 (January): 30–35.

Fahim, Hussein M. 1981. *Dams, People, and Development: The Aswan High Dam Case.* New York: Pergamon Press.

Fathy, Ali. 1976. *The High Dam and Its Impact.* Cairo: General Book.

Finley, M. I. 1999. *The Ancient Economy.* Berkeley and Los Angeles: University of California Press.

Garnsey, Peter, and Richard Saller. 1982. *The Early Principate: Augustus to Trajan.* Oxford: Clarendon Press.

Gasser, M. M., and F. El-Gamal. 1994. "Aswan High Dam: Lessons Learnt and On-going Research." *International Water Power and Dam Construction* 46, 1 (January): 35–39.

Goldsmith, Edward, and Nicholas Hildyard. 1984. *The Social and Environmental Effects of Large Dams.* San Francisco: Sierra Club Books.

Hopkins, Keith. 1988. "Roman Trade, Industry, and Labor." In *Civilization of the Ancient Mediterranean: Greece and Rome,* vol. 2, eds. Michael Grant and Rachel Kitzinger, 753–778. New York: Charles Scribner's Sons.

Hurst, H. E. 1957. *The Nile: A General Account of the River and the Utilization of Its Waters.* London: Constable.

Jacobsen, Thorkild, and Robert M. Adams. 1958. "Salt and Silt in Ancient Mesopotamian Agriculture." *Science* 128: 1251–1258.

Lanoix, J. N. 1958. "Relation between Irrigation Engineering and Bilharziasis." *Bulletin of the World Health Organization* 18.

Macmullen, Ramsay. 1988. *Corruption and the Decline of Rome.* New Haven, CT: Yale University Press.

Mansfield, Peter. 1971. *The British in Egypt.* London: Weidenfeld and Nicolson.

Morley, Neville. 1996. *Metropolis and Hinterland: The City of Rome and the Italian Economy, 200 BC–AD 200.* Cambridge: Cambridge University Press.

Pearce, Fred. 1994. "High and Dry in Aswan." *New Scientist* 142 (May 7): 28–32.

Rostovtzeff, Mikhail. 1957. *The Social and Economic History of the Roman Empire,* 2nd ed. 2 vols. Oxford: Clarendon Press.

Schiavone, Aldo. 2002. *The End of the Past: Ancient Rome and the Modern West.* Cambridge, MA: Harvard University Press.

Simkhovitch, Vladimir Grigorievitch. 1921. "Rome's Fall Reconsidered." In *Toward the Understanding of Jesus and Other Historical Studies,* 84–139. New York: Macmillan.

Smith, Norman. 1971. *A History of Dams.* London: Peter Davies.

Stanley, Daniel Jean, and Jonathan G. Wingerath. 1996. "Nile Sediment Dispersal Altered by the Aswan High Dam: The Kaolinite Trace." *Marine Geology* 133, 1/2 (July): 1–9.

Theroux, Peter. 1997. "The Imperiled Nile Delta." *National Geographic* 191, 1 (January): 2–35.

Waterbury, John. 1979. *Hydropolitics of the Nile Valley.* Syracuse, NY: Syracuse University Press.

White, Gilbert F. 1988. "The Environmental Effects of the High Dam at Aswan." *Environment* 30 (September): 4–40.

Willcocks, William. 1903. *The Nile Reservoir Dam at Assuan and After.* London: E. & F. N. Spon.

Williams, Stephen. 1985. *Diocletian and the Roman Recovery.* New York: Methuen.

IMPORTANT PEOPLE, EVENTS, AND CONCEPTS

Abbasids   A dynasty of rulers of the Muslim Empire from their capital in Baghdad (750–1258). Their rule was marked by a flowering of Qur'an studies, literature, the preservation and study of Greek learning, and medicine. A Mongol invasion destroyed the city and killed the last Abbasid ruler in 1258.

acid precipitation, acid rain   Rain bearing acidic compounds of nitrogen or sulfur derived from pollution. It is highly damaging to plants and to aquatic life.

acropolis   Greek term meaning "high city." A defensible fortress, usually on a high point within a city, possibly containing a palace or temple(s).

Akkad, Akkadian   A country and people in central Mesopotamia. Also their language, which belongs to the Semitic language group. The Babylonians were speakers of Akkadian, which therefore became the lingua franca of trade and diplomacy in the Near East during much of the second and first millennia BC.

Alberti, Leone Battista   Lived 1404–1472. Born in Venice, a humanistic philosopher, poet, painter, musician, architect, artist, and inventor. He designed an aqueduct and the Trevi Fountain in Rome for Pope Nicholas V. His many writings include the famous work on architecture, *De Re Aedificatoria*, and the lost book on shipbuilding, *Navis*.

alluvial, alluvium   Land formed by sediment, such as silt, sand, and gravel, deposited by flowing water as it slows. Most of the soil in the low-lying areas of Mesopotamia is alluvial.

Amanus   A mountain range near the Mediterranean coast of Syria, now in Turkey. It rises above the port of Iskenderun (Alexandretta). Its eastern slopes are drained by the Orontes River, which flows through Antioch.

amphitheater   A large Roman structure that could seat thousands of people to witness spectacles. The floor of the amphitheater was called an *arena*, the

Latin word for the sand that covered it in order to soak up the blood from animals and humans that were killed in combats staged for entertainment.

Anatolia   Asia Minor, corresponding to the modern area of Turkey in Asia. *Anatolia* derives from the Greek word for "east."

anoxic   Water deficient in dissolved oxygen due to biological oxygen demand, and therefore inhospitable to fish and other marine organisms.

Apennines   The mountain range that forms the backbone of Italy. Its highest peaks rise above 2,900 meters (9,500 feet).

aquaculture   An industry based on raising fish within tanks or enclosures. Practiced in the Mediterranean since ancient times, notably by Roman proprietors and in the lagoons of Venice, it produces more than 500,000 tons of fish annually, but also makes nutrient-laden waste that is often flushed into the sea.

aqueduct   A structure built to conduct water from one place to another for human use.

Arabic numerals   A system of mathematical notation, including the concepts of place value and the zero, brought from Indian sources by Arabic scholars in the eighth century.

Aragon   In northern Spain, one of the kingdoms that served as the nucleus of the nation of Spain.

Aristotle   Lived 384–322 BC. Greek philosopher who made many important observations of the natural environment. Among other aspects of nature, wrote on animals, meteorology, and physics. Speculated that teleology, or purpose, guides the phenomena of nature.

Arsenal   The naval dockyard in Venice. The term was later applied to any factory for the making of weapons. From the Arabic *dar as-sina'ah*, "house of manufacture."

Artemis   Greek goddess of everything wild, including mountains, forests, and wild animals, especially their young. Hunters worshiped her and sought her aid. Greek young women worshiped her in the form of a bear.

Assyria   A country and people located in northern Mesopotamia that rose to importance during the second and first millennia BC.

astrolabe   An instrument used to determine the altitude of the sun or other celestial bodies above the horizon, and to establish latitude and local time. Known from the time of ancient Alexandria, it was superseded in the eighteenth century by the sextant.

Aswan High Dam   Built 1964–1971. A colossal hydroelectric project on the upper Nile River in Egypt at the First Cataract. The decision to build it was made under **Gamal Abdel Nasser.** In addition to electricity, it provides flood control and water for irrigation. It has also created a multitude of environmental problems.

Atatürk   See **Kemal Atatürk, Mustafa.**

Atatürk Dam   Part of Turkey's GAP (Southeast Anatolia Development Project), this huge dam was constructed on the upper Euphrates River. One of the largest earth-and-rock-fill dams in the world, 184 meters (604 feet) high and 1,820 meters (1.13 miles) long, it was completed in 1990. Its reservoir covers 816 square kilometers (315 square miles), with a capacity greater than the Euphrates River's annual flow. Named after **Kemal Atatürk.**

Augustus Caesar (Gaius Octavius, Octavian). Lived 63 BC–AD 14, reigned 27 BC–AD 14. Founder and first emperor of the Roman Empire. Established regular taxation and principles governing trade and exploitation of resources.

autarky   Economic self-sufficiency, in which a society depends as much as possible on the resources of its own territory and on local labor.

Babylon, Babylonia   A city and country located on the Tigris River in central Mesopotamia, and the capital of the Akkadian or Babylonian Empire. An important archaeological site not far from Baghdad in modern Iraq.

barley   A grain-bearing grass widely cultivated in the early Mediterranean and Near Eastern lands.

biodiversity   "The variability among living organisms from all sources including terrestrial, marine and other aquatic ecosystems and the ecological complexes of which they are part; this includes diversity within species, between species and of ecosystems." (This definition is from the Convention on Biological Diversity, signed by more than 150 nations on June 5, 1992, at the UN Conference on Environment and Development in Rio de Janeiro.)

biomass   The total quantity of organic plant and animal material in a given area.

Black Death   Also known as the bubonic plague. A disease, *Yersinia pestis*, carried by rat fleas, which ravaged the Mediterranean basin between 1347 and 1351, killing at least one-fourth of the population. There were also later outbreaks. The plague of 542–558 under Justinian is thought by some to have been an earlier occurrence of the same disease.

Boccaccio, Giovanni   Lived 1313–1375. An Italian writer whose prolific output includes the *Decameron,* which is set in a villa in Fiesole where three youths and seven ladies have fled to escape the pestilence that is devastating the population of Florence. His work is full of information on contemporary life and on the effects of the Black Death.

Bona Dea ("The Good Goddess")   Roman goddess, daughter of Faunus (the Roman counterpart of Pan). She was worshiped by women, and her cult contains hints of wild nature, pastoralism, and wine-culture.

Braudel, Fernand   Lived 1902–1985. A French historian who stressed the importance of knowledge of the role of the natural environment in history. He was part of a movement among historians called the Annales School after the journal in which many of their writings were published. His most famous book is *The Mediterranean and the Mediterranean World in the Age of Philip II*, first published in French in 1966.

Bronze Age   The period of the early cities when tools and weapons were made of bronze, an alloy of copper and tin. Approximately 4000 to 1000 BC.

Byzantine Empire   The Eastern Roman Empire, with its capital in Constantinople. Its language and culture were Greek, and its religion was Orthodox Christianity. Early in the Middle Ages, it controlled Asia Minor, Greece and the southern Balkan Peninsula, southernmost Italy, and the islands of Cyprus, Crete, and Sicily. It lost one territory after another, ending when Constantinople finally fell to the Turks in 1453.

Cairo   The Arab capital of Egypt, called Al-Qahira, founded after the conquest in the seventh century, located at the head of the Nile Delta not far from the site of the ancient city of Memphis.

Camargue   An area in southern France near the Mediterranean that includes wetlands. In addition to horses and cattle, it is a habitat for many species of wildlife. A nature reserve was set aside there in 1927.

Canary Islands   Located outside the Strait of Gibraltar off the African coast, these islands have a Mediterranean climate. Their aboriginal inhabitants were the Guanches. They were conquered by Spain in the fifteenth century. "Canary" refers to the native dogs (*canes* in Latin), and was only later applied to the birds.

carnivorous   An animal that eats other animals, predominantly.

Carthusian   A monastic order founded in France in the eleventh century. The monks were expected to labor in useful activities including metallurgy.

cassiterite (tinstone)   A mineral, tin oxide, that is an important ore from which tin is smelted. Believed to be named after the Kassites, a people whose homeland in Elam possessed this mineral.

Castile   Called *Castilla* in Spanish, one of the kingdoms that served as the nucleus of the nation of Spain. Isabella, the patron of Christopher Columbus, was a queen of Castile.

Cato, Marcus Porcius (The Elder or the Censor)   Lived 243–149 BC. Roman politician whose work *De Agricultura* (*On Agriculture*) contains much information on the natural environment.

Cedar Forest   A famous source of timber west of Mesopotamia. It is a mythical place in the *Epic of Gilgamesh,* and a real forest of uncertain location in the histories of Sargon the Great and Naram-Sin. It is variously thought to have been in the mountain ranges of Lebanon, Amanus, or Taurus.

Ceres   Roman goddess of agriculture, identified with the Greek goddess Demeter.

cholera   An infectious epidemic disease caused by a bacillus, *Vibrio comma*. It affects the individual rapidly, producing diarrhea, cramps, vomiting, and often death, sometimes within a few hours. It is spread by contact with human bodily wastes, particularly in food or drinking water. Endemic to parts of Asia including India, it spread to the Mediterranean area several times beginning in the early nineteenth century. It can be prevented to a considerable extent by good hygiene, and by providing a safe and dependable water supply.

Christendom   The nations where Christianity is the official or dominant religion.

chryselephantine   Made of gold and ivory, usually applied to statues such as the Athena of the Parthenon and the Zeus of Olympia, both by the sculptor Phidias in the fifth century BC.

Cicero (Marcus Tullius)   Lived 104–43 BC. Roman orator and writer whose voluminous surviving works include observations of culture interacting with nature.

cinnabar   The principal ore of mercury, a red mercuric sulfide.

Cold War   The period of economic and military competition between the communist nations, led by the Soviet Union, and the capitalist nations, led by the United States, between the late 1940s and the early 1990s. In the Mediterranean area, this included civil conflicts such as that in Greece, the establishment of military bases, rivalry over projects such as the building of the Aswan High Dam in Egypt, nuclear rivalry, and the struggle for Mediterranean resources.

Columbian Exchange   The introduction of organisms from Europe, Asia, and Africa to North and South America, and the reverse, aboard European ships beginning in the fifteenth century. According to Alfred W. Crosby Jr., an American environmental historian, the Europeans brought with them a "portmanteau biota" including domestic plants, weeds, domestic animals, and microbes that caused many fatal diseases, that transformed the New World and, by weakening and killing its inhabitants, made the European conquest successful. The introduction of New World plants to the Old World was also important, as was the infection of the Old World with syphilis, the one important disease that traveled from west to east.

Columella, Lucius Junius Moderatus   Active in the first century AD. Roman author, a native of Spain. Wrote the most complete agricultural manual that survives from the ancient world.

compass   An instrument using a magnetic needle to indicate north and south directions. Invented in China, it was in use in the Mediterranean by the fifteenth century.

Compitalia   A Roman winter festival in honor of certain *lares*, shadowy ancestral gods who presided over crossroads.

condotierre (plural condottieri)   A leader of mercenary soldiers who sold their services to various lords, cities, and republics, between the thirteenth and sixteenth centuries. Their offers of service often included threats and other forms of coercion.

conservationist   A person who works for a policy of regulating the use of natural resources in order to protect nature and to ensure the continued use of resources in the future.

cork oak   An evergreen tree (*Quercus suber*) that grows in Mediterranean Europe, but especially Spain and Portugal. Its bark is used to make corks for bottles of wine and other liquids, and floats for fishing nets.

corvée labor   Forced labor exacted by landlords or governments, which might be paid or unpaid. Under Muhammad Ali in Egypt in the early nineteenth century, such labor was remunerated.

Coto Doñana (Las Marismas)   An area including wetlands, pine forests, and sand dunes at the mouth of the Guadalquivir River in southern Spain. Very rich in species, especially birds, it is a protected reserve.

cotton   A plant of the genus *Gossypium*, which produces a soft white fiber attached to its seeds. Known in India since ancient times, it became a raw material for textiles in the Mediterranean area during the Middle Ages.

cranial capacity   The volume of the skull cavity that contains the brain.

Cromer, Lord (Evelyn Baring)   Lived 1841–1917. British consul-general of Egypt, 1883–1907. He presided over the construction of the first Aswan Dam.

Crusades   A series of armed pilgrimages sanctioned by the pope with the objective of establishing western Catholic Christian rule over Jerusalem and other holy sites in the eastern Mediterranean coastlands. After initial successes including the founding of the Latin Kingdom of Jerusalem in 1100, the crusaders faced determined Islamic resistance, and Jerusalem was reconquered by the skilled Muslim general Saladin in 1187. Subsequent crusades failed to take the city. The Fourth Crusade, misdirected by the Venetians, captured Constantinople in 1204. A number of diseases, including influenza, were inadvertently spread between the eastern and western Mediterranean lands by the crusaders.

Ctesibius   Lived in the third century BC in Alexandria, Egypt. Greek mechanical genius who is said to have invented a pump with a valve, a water organ, and a water clock.

cuneiform   A writing system using marks made in clay by a stylus that is triangular (wedge-shaped) in cross section. Virtually all of the languages spoken in Mesopotamian civilizations were written in variations of this system.

Dacia   A forested province conquered by the Roman emperor Trajan, corresponding to modern Romania. The conquest is depicted in a long spiral relief on Trajan's column in Rome, the most important artistic depiction of the armament, structures, and activities of the Roman army and its environmental impacts.

Damascus   Capital of Syria during the Middle Ages, and capital of the Umayyad caliphate. Its name was given to Damascus steel, used in fine sword blades.

Dante (Durante Alighieri)   Lived 1265–1321. Florentine poet, author of the *Divine Comedy.*

DDT (dichloro-diphenyl-trichloroethane)   A persistent insecticide introduced in the 1940s and used widely in the Mediterranean area and elsewhere. It is poisonous to fish and has a disruptive effect on the breeding success of predatory birds, which are high on the food chain and therefore experience high concentrations of the chemical.

deciduous   A forest consisting of trees that lose their leaves during the coldest part of the year.

deforestation   The more or less permanent removal of the forest cover from an area.

Demeter ("Grain Mother")   Greek goddess of field crops. The great mysteries of Eleusis, which identified human life with the growth of plants, were celebrated at Eleusis in honor of her and her daughter, Kore or Persephone.

Democritus   Lived in the fifth century BC. Greek philosopher, one of the creators of the atomic theory of the universe.

desertification   The process of desiccation or drying of climate in areas that historically experience a deficiency of precipitation. This may result in the expansion of deserts, as for example the Sahara on the southern margin of the Mediterranean climatic zone, or the creation of new desert areas within the Mediterranean region. Desertification may be caused by climatic change, but the process may be exacerbated by removal of vegetation on desert margins.

Diana   A Roman goddess of wild things and women. Identified with the Greek goddess Artemis.

Diocletian   Reigned AD 284–305. Roman emperor who attempted to rationalize and control the economic life of the Roman Empire through edicts on taxation, currency, prices, and occupations. More autocratic than any emperor before him, he was also the first emperor to retire voluntarily.

Dodecanese Islands   "The twelve islands." A group of islands in the southeastern part of the Aegean Sea, including Rhodes (Rodhos), Cos (Kos), Carpathos (Karpathos), and a number of smaller islands including Castellorizo (Kastelorizo). Part of the Ottoman Empire until 1912, when they were occupied by Italy. They became part of Greece at the end of World War II in 1945.

domestication   The process of bringing animals and plants under human control, and altering their genetic makeup by artificial selection into forms amenable to human use.

Earth, Mother   Called *Gaia* or *Ge Meter* by the Greeks; *Mater Terra* or *Tellus* by the Romans. Regarded as one of the oldest of all the gods, and provider of food and fertility for humans.

ecology   The relationships of living organisms with each other and with their environments.

ecosystem   An interdependent community of organisms, including animals and plants, together with the physical, chemical, and biological environment they inhabit and with which they interact. Ecosystems can exist on a wide variety of scales.

ecotourism   Travel for the purpose of enjoying and/or learning about natural areas, including wildlife and forests.

Edict on Prices   A proclamation of the Roman emperor Diocletian that set maximum allowable prices for various commodities, services, and wages. The edict failed because it did not take account of supply and demand.

endemic   A plant or animal species confined to a comparatively limited area. Islands and mountain peaks often contain a large proportion of endemic species.

Enlil   Sumerian god of the air and king of the gods. A hero god who slew Kur (Tiamat), the monster of chaos, and made the universe of her body. Patron of the city of Nippur. His role was assumed in Babylon by that city's god Marduk, and in Assyria by Asshur.

ensi   A governor or ruler of a city-state in ancient Mesopotamia. In earlier times, this title may have implied that the ruler was the representative of a god or goddess. Under an empire, it may have designated an appointee of the emperor to govern a city or province.

entropy   The tendency of a system to move toward randomness, disorder, or chaos. This is an expression of the second law of thermodynamics.

environmental history   The study of the changing relationships through time between human societies and the natural environment, including the influences of natural forces on humans, the impacts of human activities on natural systems, and the development of human thought, science, and attitudes concerning nature.

environmentalist   A person who works for the protection of nature for its own sake, or to ensure a positive environmental quality of life for humans.

ethnography   The study of human societies that depend on nonindustrial technologies, bearing similarities to societies known through archaeology.

Eucalyptus   A genus of Australian trees with a large number of species, many of which have been planted in the Mediterranean area and propagated in plantations. The trees are aromatic and poisonous to many other species of plants. They are so widespread that they seem to be a characteristic feature of the Mediterranean landscape, but they were unknown there before the nineteenth century.

Euphrates   One of the two great rivers of Mesopotamia, located to the west of the Tigris. Its major sources are in the Taurus Mountains of modern eastern Turkey, and with the Tigris it flows into the Shatt al-Arab at the head of the Persian Gulf.

Euphrates (Tabqa) Dam   An earth-fill dam built between 1968 and 1973, the largest dam in Syria. It is 60 meters (197 feet) high and 4.6 kilometers (2.85 miles) long, and creates a reservoir, Lake Assad, extending 80 kilometers (50 miles) upriver. Its electric power generation capacity is 8.9 billion kilowatt-hours, which enables it to supply power to towns and villages east of the Euphrates. Water from the dam irrigates a quarter of a million hectares in Syria.

eutrophication   The contamination of a body of water by organic or chemical nutrients such as fertilizers. It leads to overgrowth of aquatic plants, deoxygenation, and death of many organisms such as fish and higher plants.

extinction   The disappearance of a species, either locally or throughout the world.

famine   Widespread starvation and food shortages, such as those that struck the Mediterranean area in the early fourteenth century, when an increasing population, the lack of further lands for agricultural development, and deteriorating weather caused an environmental crisis.

feddan   An Egyptian measure of land area, equivalent to 1.038 acres (0.42 hectare).

feral   Wild, especially a domestic species that has escaped or been released and subsequently become established as a wild population.

Ferdinand of Aragon   Lived 1452–1516; ruled 1474–1516. King of Aragon, whose marriage to Isabella of Castile brought together the two most powerful kingdoms in Spain. Called "the Catholic monarchs," Ferdinand and Isabella completed the reconquest of Spain from the Moors and sponsored the voyages of

Christopher Columbus that eventually revealed to the Europeans the existence of the Americas.

Florence   An inland city located in Tuscany, Italy, which was a cultural, economic, and military power in the Middle Ages and Renaissance.

footprint, environmental   The extent to which the inhabitants of cities or metropolitan nations cause ecological impacts on other areas or nations through trade and other activities.

fossil fuels   Materials from organic deposits found in the earth, including coal, petroleum, and natural gas. Rich in carbon and hydrogen, these are combustible and have served as the major fuels from the nineteenth century onward.

Fracastoro, Girolamo   Lived 1478–1553. Italian physician, born in Verona, who studied at the University of Padua. He described and first named syphilis, giving the word a fantastic origin from Syphilus, the shepherd of King Alcithous of Ophyre [Haiti] who was slain by Apollo. He also described the treatment of the disease by guiacum, a derivative of guiac, a sacred tree originating in the West Indies.

French Forest Ordinance of 1669   A law enacted under French king Louis XIV at the recommendation of his minister, Charles Colbert, intended to safeguard the supply of timber for the navy by limiting other forest uses such as charcoal burning and barrel making. It placed the forests under the control of the state.

Galileo Galilei   Lived 1564–1642. Florentine scientist whose research with the telescope, including the discovery of the moons of Jupiter, led him to conclude that the Earth revolves around the sun, as Nicolaus Copernicus had earlier maintained. He was forced by the papal Inquisition to renounce his ideas in 1633. He also speculated on the causes of the tides and other environmental phenomena.

garigue   A Mediterranean plant association of low scattered bushes, many of them spiny and/or aromatic. Also called "rock heath." In many if not most cases, garigue results from the degeneration of maquis due to the impact of humans, including fire and grazing animals such as goats.

Genoa   A city located on the northwestern Italian coast that wielded great economic and naval power in the Mediterranean basin, including the Black Sea, during the Middle Ages.

Gibraltar   A fortified rock about 425 meters (1,400 feet) high that forms a peninsula on the south-central coast of Spain, and on the north side of the strait that bears the same name. It is the only European habitat of the Barbary ape. The Strait of Gibraltar is the only connection between the

Mediterranean Sea and the Atlantic Ocean. Gibraltar is a British Crown Colony but is claimed by Spain.

Gilgamesh Legendary king of Uruk (Erech) in ancient Sumeria. The *Epic of Gilgamesh*, perhaps the oldest long work of literature that survives, recounts his adventures in search of forest resources as well as eternal life.

global warming An increase in accumulated heat in the Earth's atmosphere. Scientists believe it is caused to some extent by the "greenhouse effect," the tendency of certain gases such as carbon dioxide and methane to reduce the loss of heat to space once it has entered Earth's atmosphere from the sun. These gases are produced by combustion of fossil fuels and other human activities. The effects of global warming include shifts in weather patterns and a rise in average sea level due to expansion of seawater and melting of glaciers, especially on Greenland.

grapevine The plant, *Vitis vinifera*, that bears grapes used to make wine. Often called simply "vine." With grain and olives, one of the three staple foods making up the "Mediterranean triad" in agriculture.

Guanches The native inhabitants of the Canary Islands, conquered by Spaniards in the fifteenth century.

gunpowder A mixture of charcoal, sulfur, and potassium nitrate, which will produce an explosion when ignited in a confined space. Invented in China, it came to Europe by way of the Arabs around 1250. Its use as a propellant in cannons aided the Turkish capture of Constantinople in 1453, and made city walls relatively useless as a defense.

habitat An area providing a set of environmental conditions necessary for a species or community to live there.

Hammurapi (Hammurabi) King of Babylon, ruled 1792–1749 BC. He established a Babylonian empire throughout Mesopotamia and promulgated a famous law code.

hamsin A seasonal wind, usually in spring, that transports dust and sand from the desert. The name derives from the Arabic word for "fifty," since it supposedly can blow for fifty days.

Hera (Roman Juno) Greek goddess of women, queen of the gods, sister and wife of Zeus (Jupiter).

herbivorous An animal that eats plants, predominantly.

Hercules (Heracles), Pillars of The Strait of Gibraltar, the opening between the Mediterranean Sea and the Atlantic Ocean. It was the only connection of the Mediterranean Sea with any part of the world ocean before the construction of the Suez Canal. It is 13 kilometers (8 miles) wide at its narrowest point, where its depth is 366 meters (1,200 feet).

Hero (Heron) of Alexandria   Lived in the first century AD. Greek inventor and mathematician. He described machines for lifting weights, automata, odometers, a wind-driven machine, and a prototypical steam engine called an aeolipile.

Herodotus   Lived ca. 484–428 BC. Greek historian of the Persian Wars. His intense interest in the ways of life of many peoples led him to record much of interest about human relationship to the natural environment.

Iberia   The peninsula that includes Spain and Portugal.

ibis   A long-billed waterbird. The sacred ibis, formerly common in Egypt, was associated with Thoth, the god of healing and learning.

Ice Age   A variable climatic period, with conditions generally colder than in modern times, during which glaciers occupied large areas in northern lands. In the Mediterranean basin, the glaciers generally were limited to mountains. Warming trends ended the most recent Ice Age around 13,000 to 12,000 years ago.

Industrial Revolution   The modern period of extensive mechanization of production systems and the development of large-scale factory production, beginning around 1750 in England and spreading to the Mediterranean area during the nineteenth and twentieth centuries. After an initial period when it was powered by water and wood, the Industrial Revolution was dominated by coal as a fuel and by production of iron and steel.

insectivorous   An animal that eats insects, predominantly.

insula   In Rome and other Roman cities, a block of apartment buildings. The word means "island," probably referring to the fact that the block was surrounded by streets.

Ionian Islands   The archipelago west of the Greek mainland consisting of Corcyra (Kerkira), Leucas (Lefkas), Ithaca (Ithaki), Cephalonia (Kefalinia), Zante (Zakynthos), and other smaller islands. Cythera (Kithira), south of Greece, was sometimes included in this group. In the Middle Ages they belonged to Venice or sometimes Turkey. They were independent from 1800 to 1807, then occupied by Napoleon, and became a British protectorate in 1815. In 1864 they became part of Greece. They experienced major tourist development in the twentieth century, and Zakynthos became the scene of struggles between such development and the preservation of the sea turtle population.

Iron Age   In the fourteenth century BC, the Hittites, a people of Asia Minor (modern Turkey), began to make tools and weapons of iron. By around 1000 BC, this technology had spread widely, and iron replaced bronze as the dominant metal throughout the Mediterranean basin and Near East.

irrigation   Conducting water to agricultural land for crops. Mesopotamian civilization was dependent on irrigation for its food supply. In Egypt, basin irrigation used the annual flood of the Nile River, whereas perennial irrigation used artificial methods to provide water throughout the year.

Istanbul   Called Constantinople until 1453, this city had been the capital of the Byzantine Empire. After its conquest by the Turks under Mehmed II, it became the capital of the Ottoman Empire.

Jonglei Canal   A proposed project to channelize the White Nile through the vast Sudd marshes in the Sudan in order to bring water downstream and avoid loss of water by evaporation. It would destroy millions of acres of wildlife habitat. Long-term civil war in the Sudan has so far prevented its completion.

Jordan River   A river that flows 320 kilometers (198.4 miles) from northern Israel, with tributaries in Syria, to the Sea of Galilee, and thence through the nation of Jordan and the West Bank to the Dead Sea. Its waters are a matter of importance and controversy to the nations of the area.

Jupiter (Juppiter)   Roman father-god, identified with the Greek god Zeus, the power behind all weather. With Juno his consort and Minerva (a warrior-goddess), the most honored of Roman gods.

Justinian   Lived 483–565, reigned 527–565. Roman emperor who ruled from the eastern capital of Constantinople. Noted for his reconquest of much of the western part of the Roman Empire, for rebuilding the city of Constantinople after riots and fires with great structures including the church of Ayia Sophia, and for the codification of Roman law.

Juturna   A water nymph who had a shrine at a spring in Rome. Jupiter was said to have made her the goddess of lakes and streams.

Kemal Atatürk, Mustafa   Lived 1881–1938; president of the Turkish Republic, 1923–1938. He led Turkish resistance to an Allied invasion after the fall of the Ottoman Empire in World War I and turned Turkey into the first modern secular state in the Islamic world. Among his reforms were European-style laws and dress, use of the Roman instead of the Arabic alphabet, and political, legal, and social rights for women.

Kurds   A numerous people inhabiting Kurdistan, a region in southeast Turkey, northern Iraq, and western Iran. The Kurds speak an Indo-European language distantly related to Farsi, the language of Iran. Their national aspirations were frustrated by the settlements imposed by the Western Allies at the end of World War I, and they continue to cause difficulties and unrest within and among Turkey, Iraq, and Iran, all three of which oppose the creation of an independent Kurdistan.

Lake Nasser   The reservoir impounded on the Nile River by the Aswan High Dam. It extends upstream into Sudan, where it is called Lake Nubia.

lar (plural *lares*)   Minor Roman gods who presided over the countryside, farmhouses, roads, and the sea. They were supposed to have existed in the distant past as human beings.

latifundium (plural *latifundia*)   Latin term for a large ranch devoted to raising cattle and other herd animals for sale.

lead poisoning   An illness caused by absorption of lead into the human body, usually through the digestive system or the lungs. The symptoms may include loss of energy, of mental acuity, and of reproductive fertility, and the condition may result in death.

Leonardo da Vinci   Lived 1452–1519. Italian artist, sculptor, and inventor of multiple talents. He believed that a thorough knowledge of nature was necessary to a painter, and he portrayed plants and moving animals, including horses, cats, and crabs, with exactitude. He studied the movements of water and the effects of floods. His technological speculations foreshadowed the effects that human efforts would come to have on the natural environment.

Lesseps, Ferdinand de   Lived 1805–1894. French diplomat, financier, and planner. Builder of the Suez Canal, 1859–1869, at the invitation of Said Pasha, viceroy of Egypt. He also tried to build a canal across Panama, 1879–1888, but failed.

Levant   The lands bordering the eastern shores of the Mediterranean in what are now Palestine, Israel, Jordan, Lebanon, western Syria, and extreme southeastern coastal Turkey.

littoral   A shore or coastal region. Sometimes used to indicate to what extent a region of land is contiguous to, or surrounded by, the sea.

López de Villalobos, Francisco   Lived 1473–1556. A Spanish professor of medicine at the University of Salamanca, he was the first doctor in Spain to write about syphilis, 1493–1495, and to attribute its arrival in Spain to the sailors who returned with Columbus from the West Indies, where they had supposedly contracted the disease from Carib women.

Lowdermilk, Walter Clay   Lived 1888–1974. American soil conservationist and forester who traveled in the Mediterranean area and wrote about agriculture and soil conservation there in two books, *Palestine: Land of Promise* (1944) and the widely distributed *Conquest of the Land through 7,000 Years* (1953). He served as an agricultural advisor in China and Israel.

Lucretius, Titus Lucretius Carus   Lived in the first century BC. Author of a philosophical poem, *De Rerum Natura*, which attempts to describe the physical universe. He embraced the atomic theory.

Macaronesia   Greek term meaning "Islands of the Blessed" or "Blessed Islands." Refers to the Canary Islands, Madeira, and the Azores.

Machiavelli, Niccolò   Lived 1469–1527. Florentine statesman, author of *The Prince, The Art of War,* and *The History of Florence.* Best known for his realistic view of politics and his rejection of conventional morality, he also speculated on the environmental conditions that guarantee the survival and success of a state.

Madeira   A group of islands in the Atlantic Ocean outside the Strait of Gibraltar. *Madeira* is Portuguese for "wood"; the main island was heavily wooded and uninhabited when the Portuguese discovered it in the fifteenth century. Christopher Columbus lived in Madeira before his voyages across the Atlantic.

Maghreb   The mountainous and coastal area of western North Africa, including Morocco, northern Algeria, and Tunisia. *Maghreb* means "west" in Arabic. This area includes a coastal lowland along the Mediterranean, the Atlas Mountains, and the northern fringe of the Sahara. Originally well vegetated, the Maghreb has undergone a process of desertification and deforestation throughout recorded history.

maize   Usually called corn in the United States. A domesticated food grass that originated in the Valley of Mexico, maize was one of the staple foods of the pre-Columbian Americas, and after its introduction to the Old World became one of the major food crops of humankind.

malaria   A severe chronic infectious disease transmitted by the bite of the female *Anopheles* mosquito, which breeds in water. Characterized by fever and pain, it often eventuates in death. The word in Italian means "bad air," since in the Middle Ages it was believed that it was caused by the mephitic atmosphere of marshes.

manor   The land and its human and economic resources that were under the control and management of a single lord. The basic unit of the medieval agrarian economy.

Manutius, Aldus   Lived 1449–1515. A great Venetian scholar-printer who published many books, including classical Greek texts. He originated Italic font letters.

maquis   A typical Mediterranean evergreen plant association of shrubs and low-growing trees characterized by resistance to drought. It is subject to periodic fires, but is adapted to recover rapidly after them.

Marcus Aurelius   Lived 121–180, reigned 161–180. Roman emperor and Stoic philosopher, author of the *Meditations,* written in Greek. A great plague struck the Roman Empire during his reign.

Marsh, George Perkins   Lived 1801–1882. A scholar, statesman, and diplomat who served as U.S. ambassador to Turkey, and to Italy from 1861 to 1882. He wrote an influential book, *Man and Nature* (1864), which described the effects of human activities on the natural environment, both positive and negative. In particular, he held that human actions tend to degrade the environment. He is often regarded as the first environmental historian.

mechanization   The change to use of machinery in human activities that formerly depended on human and animal labor, such as agriculture. Generally mechanization increased the impact of these activities on the environment.

Mediterranean   The term *Mare Mediterraneum*, first used by the Roman writer Gaius Julius Solinus in the third century AD, means "the Mid-Earth Sea," or "the Sea in the Middle of the Earth." It was also called by the Romans *Mare Internum* (or *Intestinum*), "the Inner Sea," and *Mare Nostrum*, "Our Sea."

Mediterranean Action Plan (MAP)   An agreement among the Mediterranean coastal states, negotiated in 1975, to gather information about pollution and to set standards to control it.

Mercury   Roman god associated with pastoralism. Patron of shepherds, travelers, and thieves. Identified with the Greek Hermes, the messenger of the gods.

mercury   A metal, liquid at ordinary temperatures, that may damage the nervous system of vertebrates, including humans. Its organic compounds are highly poisonous. It is used extensively in the chemical and plastics industries, in processing gold, in dentistry, and in detonators. It is an extremely dangerous pollutant in the Mediterranean region.

merino   A sheep of a breed producing exceptionally fine wool suitable for knitting apparel. Originating among the Berbers in Morocco, it was brought to Spain by the Muslims and became the most valued breed there from that time forward. It is regarded as more damaging to vegetation than other breeds of sheep because it tends to exhaust the grass in one place before moving on.

Mesopotamia   "Land between the rivers." The region watered by the Tigris and Euphrates rivers, which saw the rise of the Sumerian, Akkadian-Babylonian, and Assyrian civilizations. It occupied the area now included in the nations of Iraq and Kuwait and parts of Iran, Turkey, and Syria.

Mesta   A society formed in the thirteenth century by aristocratic sheep owners in Spain to represent their interests at court. They gained the right to use and control the trackways, or *cañadas*, along which the sheep were driven during transhumance, over hundreds of miles across Spain. Eventually the king became Grand Master of the Mesta, and its power was unchallenged until the nineteenth century.

metallurgy   The technology of extracting metals from their ores.

Miocene epoch   One of the time divisions of the Cenozoic era, which lasted from about twenty-five million years ago to five million years ago.

moldboard   A heavy plow with a curved plate that turns over the furrow slice.

Monaco   An independent principality on the French Mediterranean coast, 150 hectares (370 acres) in area. It is the site of an oceanographic institute.

monastery   A house of an order of monks or nuns. These communities engaged in many types of labor on the land, becoming important centers of production and a force of development in mountainous and other marginal areas.

monk seal (*Monachus monachus*)   An aquatic mammal that formerly lived throughout the Mediterranean and Black seas and in the Atlantic on the West African coast and the Macaronesian Islands. Today it is seriously endangered, with a few isolated groups surviving in the eastern Mediterranean and the Atlantic.

Mongols   A central Asiatic people whose invasions in the thirteenth century under Genghis Khan and his successors reached Mesopotamia, Syria, and the Black Sea area.

monoculture   The planting of a single crop species or species of tree over a very large area. Monoculture increases the vulnerability of a crop to pests and diseases, which can spread easily from one susceptible plant to another.

Monte Circeo   A varied area of Mediterranean coastline on the Tyrrhenian coast of Italy. Rich in forests and wildlife, it was declared a national park in 1934.

Moriscos   Spanish people of Moorish (Islamic Moroccan) descent, nominal converts to Christianity. Many of them were experts in irrigation agriculture and were affluent. They were expelled from Spain by the king with great cruelty in 1609.

Mount Olympus   One of the highest peaks in the Mediterranean area, at 2,911 meters (9,606 feet) in elevation, it was considered to be the home of the greatest ancient Greek gods. It is a national park of Greece.

Muhammad   Lived 570–632. The greatest prophet of Islam, he lived in Mecca and Medina, Arabia.

Muhammad Ali   Lived 1769–1849. Viceroy of Egypt, nominally under the Ottoman Empire, 1805–1848. A military man of Albanian descent, he modernized the Egyptian economy and took steps toward making it part of the world market economy. He began some works to control the Nile and to develop artificial irrigation allowing continuous cropping.

Napoleon I   Lived 1769–1821. Born on the Mediterranean island of Corsica, he later ruled as emperor of the French, 1804–1815. In addition to France, his conquests in the Mediterranean region included Spain, Italy, Illyria, and Malta. He invaded Egypt, taking Alexandria on July 1, 1798, and entering Cairo on July 24. After a campaign in Palestine and Syria, he abandoned his troops in Egypt on August 24, 1799, and returned to France. During his brief stay, French scientists carefully studied the peoples and the environment of Egypt, including its animals and plants, and made valuable reports and drawings of the ancient monuments, all of which were published between 1809 and 1828 in a series of volumes entitled *Description de l'Égypte* (*A Description of Egypt*).

Naram-Sin   King of the Akkadian Empire, ruled 2254–2218 BC. Grandson of Sargon the Great. He made many conquests and claimed divine kingship. He secured the road for timber imports from the western mountains.

Nasser, Gamal Abdel   Lived 1918–1971. Leader of the revolution that established Egypt's independence in 1952, later prime minister, and then president of the United Arab Republic. He fostered the building of the Aswan High Dam and worked to transform Egypt into an industrial nation.

National Museum of Iraq   Located in Baghdad, this is one of the most important museums of ancient art, archaeology, and history in the world. It contained tens of thousands of objects recovered from archaeological sites in Iraq, including major works of art and collections of cuneiform tablets, as well as the records pertaining to their provenance. The museum was pillaged and vandalized, and records scattered and destroyed, by looters in the early days of the U.S. occupation of Baghdad in April 2003. Some objects have been recovered, and it is hoped that more will be found, but the loss remains regrettable.

national park   An area delimited and protected by a state to preserve important natural or historical features. Most Mediterranean nations have established national parks.

National School of Waters and Forests (Eaux et Forêts), Nancy, France   The world's first great forestry school, founded in 1824, with influences that transformed forestry around the world, including the Mediterranean lands, especially southern France. The idea of sustainable forest management was one of the most important principles of the Nancy school.

Natufian   A society that depended on intensive gathering of wild plants between 13,000 and 10,000 years ago in the Levant. It is named after Wadi en-Natuf, the site in Palestine where it was first discovered. Natufians made sickles and various tools to grind grain.

Neanderthal   A human subspecies or species that was widespread in Europe and western Asia between about 70,000 and 30,000 years ago. They were stocky and muscular, with pronounced brow ridges, and were hunter-gatherers. They made tools and works of art, and practiced burial of the dead with grave offerings. It is not known whether they were ancestors of modern humans.

Nile River   A river that rises in East Africa, flows through Egypt, and reaches the Mediterranean Sea through mouths in the delta. It is Egypt's only considerable source of water. At approximately 6,400 kilometers (4,000 miles) in length, it is the longest river in the world.

nilometer   A device to measure the height of the annual flood of the Nile River, it consists of a shaft with depth markings that connects to the river.

nonrenewable resource   A resource that exists in a fixed quantity, at least within the time scale of human lives and generations. Examples are metallic ores and fossil fuels such as coal, oil, and gas. Such a resource will be used up in a period depending on the rate of use. The period can be extended by recycling.

Nubia   A region on the Nile River above Aswan in Egypt and the Sudan. It was largely flooded by the Aswan High Dam and its inhabitants resettled elsewhere.

***numen* (plural *numina*)**   Roman term for a supernatural power or influence, often identified with various aspects of nature.

olive   The tree *Olea europaea* and its edible fruit, the source of oil. One of the three basic food crops of the Mediterranean region, along with grains and the grapevine.

opium poppy   The plant *Papaver somniferum*, the source of opium and narcotics derived from it. One of the most important medicinal plants in Mediterranean history.

Ottoman Empire   The great Turkish empire that dominated half of the Mediterranean basin during the early modern period. It was founded by Osman (also called Othman, 1290–1326) in northwestern Asia Minor. The sultan Mehmed II captured Constantinople in 1453, renamed it Istanbul, and made it the Ottoman capital. The empire lasted until World War I.

palynology   The study of ancient plant pollen found in various deposits, such as those in lake bottoms and caves, to determine what kinds of vegetation formerly existed, and how they changed.

pandemic   An epidemic of contagious disease that affects a large portion of the human population in an extensive region.

paradise   An enclosed natural or gardenlike area reserved for worship of the gods, or for hunting by kings and other royalty. Derives from a Persian word denoting a place surrounded by a wall.

pastoralism   A way of life in which people care for, subsist upon, and control the movements of herds of large herbivorous animals, using their products such as meat, milk, skins, and wool. In the Mediterranean area the animals involved in pastoralism are most notably sheep, goats, cattle, horses, and camels.

patrician   A member of the Roman upper class of citizens, defined by descent from a noble family.

peasant   A member of the class of laborers on the land, subject to the manor lord in the feudal system.

penates   Roman gods that represented natural powers, whose mysterious action produced everything that humans cannot accomplish for themselves, but which is necessary to life and prosperity.

photochemical   Chemical reactions whose energy is supplied by sunlight. Many of the reactions between the various constituents of smog are of this type.

phylloxera   An aphid insect that is very destructive to grapevines. Native to the eastern United States, phylloxera was observed in France in 1863 and spread to virtually every wine-producing district in France and around the Mediterranean. As a defensive measure, plant scientists successfully grafted the European grape varieties onto American rootstock, and in 1881 a conference of French winegrowers approved the replacement of susceptible vines with the new grafted ones, rescuing the production of wine. Similar methods saved vineyards all around the Mediterranean, including Spain, the Maghreb, the Balkans, and Greece.

phytoplankton   Plant organisms that live near the surface of the sea and that drift with the current.

Piccolomini, Aeneas Silvius   Lived 1405–1464, and served as Pope Pius II, 1458–1464. He grew up on his father's farm and developed an enthusiasm for the natural environment. Educated at Siena, he became a poet, novelist, playwright, and naturalist.

Pico della Mirandola, Giovanni   Lived 1463–1494. An Italian humanist who raised many controversial questions and attracted the suspicion of the pope, although he was never condemned. He studied the Hebrew Kabbalah, doubted astrology, and affirmed the kinship of humans with animals.

piling, pile   A heavy timber driven into the soil as a foundation or support for a building. For example, Venice was supported on thousands of pilings.

Pinus radiata   Called Monterey pine in the small region of coastal California where it is native, this fast-growing species has been established in plantations in areas of Mediterranean climate, or similar climates, around the world. It is often seen in the Mediterranean basin.

Pirenne, Henri   Lived 1862–1935. Belgian historian, author of influential works including *Muhammad and Charlemagne*, which argued that the history of the European Middle Ages can only be understood by studying the Muslim Middle East.

Pisa   A city in Tuscany, Italy, that had access to the sea by way of the navigable lower Arno River. Pisa was a major center of sea trade until it was conquered by Florence in 1410 and made that city's port.

plate tectonics   The movement of large sections of Earth's rocky crust relative to one another, driven by the upwelling of magma (fluid molten rock) from a layer underneath.

Plato   Lived ca. 429–347 BC. Greek philosopher. In some of his dialogues, particularly the *Critias* and the *Laws*, there are observations on human interactions with the natural environment, including deforestation and use of resources such as water.

plebeian   A member of the ordinary or common class of Roman citizens.

Pliny, Gaius Plinius Secundus, the Elder   Lived AD 23–79. Roman writer of a series of books on natural history, which are a collection of descriptions and reports of almost everything then known or imagined in the world of nature, both true and fallacious.

Plutarch of Chaeronea   Lived AD 50–120. Greek biographer in the time of the Roman Empire, he wrote a series of lives of famous Greek and Roman men, often comparing one Greek figure with a Roman counterpart, such as Alexander the Great and Julius Caesar.

Po River (*Padus* in Latin)   The great river of northern Italy, which flows about 670 kilometers (415 miles) from the Alps to the Adriatic Sea south of Venice.

poacher   A person who hunts, gathers, or fishes illegally.

polenta   A Mediterranean food made from ground maize meal, especially popular in northern Italy.

poliomyelitis   Often called polio or infantile paralysis. A highly infectious disease that can be spread through contaminated food or drinking water. It most commonly affects children and young adults. It can result in paralysis of the limbs and the muscles of the respiratory system, in some cases resulting in death. A vaccine provides complete protection against the disease, and has virtually eliminated it in the Mediterranean region and other parts of the developed world.

pollution   The alteration of any part of the environment, such as water, air, or soil, so as to make it hazardous to human health. Often refers to the discharge of chemicals or sewage, or the venting of smoke and fumes to the atmosphere.

polychlorinated biphenyls (PCBs)   A group of artificial chemical compounds created as a result of certain industrial processes. They are known to be cancer-causing to humans and other animals.

Pontine (Pomptine) Marshes   An area of wetlands and rich soil south of Rome, well-known as a center of malarial fevers and a place where locust infestations bred. There were several attempts to drain the area by canals and other works down through history, beginning with the censor Appius Claudius in 312 BC (his construction of the Appian Way, however, may have made conditions worse) and including Julius Caesar, who was assassinated before his plans could reach fruition.

printing   Block printing is known from China in the ninth century, and movable type, where each character was carved on a separate block, from the eleventh century. It came into use in Europe and the Mediterranean in the fifteenth century in the form of movable leaden type cast in molds.

putting-out   A system of manufacture in which a merchant advances to artisans the raw materials and money for wages, and then sells the finished product.

reclamation   Constructing irrigation or drainage works to render uncultivated land suitable for agricultural use.

renewable resource   A resource that is constantly replenished, so that it can be used sustainably. Examples are forests, fish, and other animal and plant populations. A renewable resource can be depleted or destroyed if the rate of use is greater than the rate of replacement. Sometimes the meaning is extended to forms of energy such as solar radiation, wind, waves, and tides.

salinization   The accumulation of salts in the soil when water evaporates. It often occurs in irrigated but poorly drained soils in warm climates. Salts can reach concentrations that are toxic to plants.

Sargon I the Great   King of the Akkadian Empire, ruled 2334–2279 BC. Also called Sargon of Akkad (Agade). The founder and ruler of the first Mesopotamian empire.

Saturn   Roman god of crops and agricultural land.

schistosomiasis (bilharzia)   A disease caused by parasitic trematode worms that pass into water in urine and feces and infect snails as alternate hosts. The disease is very debilitating and can lead to death. It is common in Egypt and other warm countries where people work in contact with water for long periods of time.

sea turtles   Five species of sea turtles are found in the Mediterranean, and all are endangered. The loggerhead and green turtles live, feed, and nest in the Mediterranean and are most in danger from disturbance and destruction of their nests on beaches. Hawksbill, leatherback, and Kemp's Ridley enter the Mediterranean from the Atlantic. All turtles are in danger from pollution, rubbish, collision with boats and their screws, and the nets and hooks used in the fishing industry.

Seneca, Lucius Annaeus (the Younger)   Lived 4 BC–AD 65. Roman philosopher who made many observations on human activities relating to nature.

sharecropping   A system of land tenure in which a lord allowed the use of land, tools, and draft animals, and in return received a proportion of the harvest.

Shatt al-Arab   A strategic tidewater estuary stretching from the mouths of the combined Tigris and Euphrates rivers to the Persian Gulf. A portion of it forms part of the boundary between Iraq and Iran.

Shulgi   Ruled 2097–2094 BC. Son of Ur-Nammu, king of the Third Dynasty of Ur. He established a law code that regularized the economy of Mesopotamia.

Silk Road   A series of trade routes across inland Asia connecting China with the Mediterranean area.

silphium   A medicinal plant, now extinct, a member of the umbellifers or carrot family, which formerly grew in the Cyrenaica region of Libya.

silt   A sedimentary material, often carried by rivers, consisting of small sandy or muddy particles.

Silvanus   Roman god of woodlands and groves of trees, later identified with Faunus, another Roman deity, and with the Greek god Pan.

sirocco   A hot wind blowing northward from the Sahara. Dry and dusty at first, it picks up moisture over the Mediterranean Sea and causes uncomfortable weather on the northern side of that sea. By pushing the water of the Adriatic Sea northward, it can cause floods in Venice.

slavery   A system of ownership of some humans by others, practiced in all ancient Mediterranean societies. A slave under Roman law was *instrumentum vocale*, a speaking tool. Slavery contributed to environmental damage in part by depriving its subjects of motivation for environmental preservation and improvement.

smallpox   A highly infectious and often fatal disease that produces fever, aches, and a widespread eruption that leaves pockmarks on victims that survive. It is one of the most important epidemic diseases in Mediterranean history. It is caused by a virus and is spread by contact with infected persons or objects with which they have been in contact. Vaccination with cowpox or serum from mammals that have immunity to smallpox is very effective in preventing the disease and has resulted in its virtual eradication.

smelting   The process of separating metals from their ores by melting or fusing at high temperatures.

soil exhaustion   The loss of nutrients in the soil that are essential to vigorous growth of food plants. It may be caused by overuse of agricultural soil and can be counteracted by fertilization or planting of nitrogen-fixing crops such as legumes.

Southeast Anatolia Development Project   Also called GAP, an acronym based on the Turkish-language name. A plan drafted by the Turkish government to utilize the waters of the Tigris and Euphrates rivers for hydroelectric power generation and irrigation, including the construction of some twenty-two dams and nineteen power plants. Its largest unit is Atatürk Dam, one of the largest earth-and-rock-fill dams in the world, 184 meters (604 feet) high and 1,820 meters (1.13 miles) long. Its reservoir covers 816 square kilometers (315 square miles), with a capacity greater than the Euphrates's annual flow.

Spice Islands   The Moluccas and other islands in Indonesia that supplied cloves, cinnamon, and other spices to the Mediterranean area beginning in the fifteenth century.

Suez Canal   An artificial waterway in Egypt permitting ship traffic between the Mediterranean Sea and the Red Sea. It also permits the movement of fish and other aquatic organisms between the two seas, but most of such movement has been into the Mediterranean. Constructed by the French engineer and entrepreneur Ferdinand de Lesseps between 1856 and 1869, it is about 160 kilometers (100 miles) long, and greatly reduces the distance and travel time by sea between Europe and Asia.

Sumer, Sumeria   A civilization, one of the first on Earth, located in the southern part of Mesopotamia during the fourth and third millennia BC.

sumptuary laws   Laws limiting the display of wealth or luxury, such as the wearing of golden ornaments or purple cloth, or the erection of expensive funeral monuments.

syphilis   A chronic venereal disease caused by a spirochete, *Treponema pallidum*, believed to have been brought from the New World to the Mediterranean area by sailors, possibly for the first time by those who sailed with Christopher Columbus in 1492.

Taurus   A mountain range that marks the northern margin of Mesopotamia. Running generally east and west, it parallels the Mediterranean coast of modern Turkey and thence eastward into northwestern Iran.

technology   The knowledge and skills involved in collecting or extracting materials, fashioning implements, and using them to carry on human activities.

terraces   Steplike strips constructed for agricultural purposes, including erosion control, often with stone walls for support.

Tertiary period   The first part of the Cenozoic era. It lasted from about 65 million to 2 million years ago. It consists of the Paleocene, Eocene, Oligocene, Miocene, and Pliocene epochs.

Tethys Ocean   A very large body of water that existed in past geological epochs between Africa and Eurasia. As it narrowed due to plate tectonics, the Mediterranean Sea was formed.

Tharthar   A depression occupied by a lake between the Euphrates and Tigris rivers in Iraq north of Baghdad. Canals and other hydraulic works allow a portion of the waters of the Tigris to be diverted through Tharthar Lake to the Euphrates, providing a measure of flood prevention for the city, and allowing regulation of waters for irrigation in the lower Euphrates basin.

theophany   An appearance or revelation of God in something visible, such as creation, a creature, or a human being.

Theophrastus   Lived ca. 372–287 BC. Greek philosopher, student of Aristotle. He wrote on many environmental subjects, including plants, stones, and winds, and made ecological observations.

Third Dynasty of Ur   A Sumerian empire (2212–2004 BC) that fostered a renaissance of Sumerian culture and literature.

three-field system   An agricultural regime in which crops are planted and harvested for two years, and the land is allowed to lie fallow during the third year. This system, more intensive than the **two-field system**, became widespread in Europe, especially north of the Alps, in the Middle Ages.

Tiamat (Sumerian Kur)   A chaotic water-monster, defeated by the hero-god Enlil, who split her body in two, using one half to make the sky and the other the earth.

Tigris   One of the two great rivers of Mesopotamia, located to the east of the Euphrates. Its major sources are in the eastern part of modern Turkey and in the Zagros Mountains of Iran, and with the Euphrates it flows into the head of the Persian Gulf.

tourism   Travel for pleasure, and the industry based on providing services for tourists. Tourism has become one of the leading industries in many Mediterranean countries.

Tours, Battle of   Near the city of the same name in west-central France, Charles Martel defeated a Muslim army in 732. The battle marked the furthest advance of the Muslim military into western Europe.

Trajan   Lived AD 53–117, reigned 98–117. Roman emperor under whom the empire reached its widest territorial extent. His edicts demonstrated concern with the economic welfare of the Roman people, including one that required landowners receiving government grants to make subsistence payments to poor children in rural districts.

transhumance   The movement of herders with their livestock to different locations in different seasons. Often this means going to mountain pastures in the summer and warmer lowlands in the winter. In Spain, the movement was also to the north in summer and the south in winter.

trawler   A fishing boat that drags a large fishing net along the sea bottom and/ or uses a fishing setline with multiple hooks. Both methods destroy many marine organisms besides the target species.

trophic   The characteristic diet of an animal species, for example herbivorous, carnivorous, etc.

two-field system   An agricultural regime in which a crop is planted and harvested, and the land left fallow during the alternate year. This system was characteristic of the Mediterranean lands in ancient and medieval times.

typhus   A rickettsial disease transmitted to humans by lice that infest both rats and humans. Its spread is exacerbated by environmental conditions such as overcrowding and lack of sanitation. Today it can be treated by antibiotics.

Umayyad   A dynasty of rulers of the Muslim Empire from their capital in Damascus (661–750), and of Muslim Spain, ruling in Cordoba (758–1002).

United Nations Environment Programme (UNEP)   The UN agency that coordinates many efforts of member nations and other agencies in certain areas of environmental conservation. It was organized in 1973 as a result of the UN Conference on the Human Environment in Stockholm, held the previous year. Its world headquarters are in Nairobi, Kenya.

United Nations Special Commission (UNSCOM)   Established in April 1991 by Security Council Resolution 687 and given the responsibility to inspect and supervise the destruction of Iraq's weapons of mass destruction. The work of the commission ended in October 1998, when Iraq ended all forms of cooperation with it. A successor body, the United Nations Monitoring, Verification, and Inspection Commission (UNMOVIC), was created in December 1999 by UN Resolution 1284, but was rejected by Iraq. Iraq accepted UNMOVIC in November 2002, but UN Secretary-General Kofi Annan withdrew the inspectors in March 2003, responding to the statement by the U.S.-led coalition that war was imminent.

Ur (modern Tell al-Muqayyar)   A prosperous city of early Sumeria, it became capital of the powerful empire of the Third Dynasty of Ur (2112–2004 BC).

urbanization   The origin and growth of cities as areas of human habitation, including transportation, markets, government and religious buildings, and other infrastructure.

Ur-Nammu   Ruled 2112–2095 BC. First king of the Third Dynasty of Ur, he claimed the title of king of Sumer and Akkad. He constructed many temples and other monumental structures.

Uruk (Erech, Warka)   The Sumerian city of Gilgamesh, who according to legend built its massive walls. The most important city in Mesopotamia in the fourth millennium BC and the location of several major temples. The famed Warka vase, carved in 3000 BC or earlier with scenes of agriculture and domestic economic activity, part of a temple treasury hoard, was lost in the looting of the **National Museum of Iraq** in Baghdad in 2003.

Varro, Marcus Terentius   Lived 116–27 BC. Roman author of a treatise on agriculture, cattle and sheep breeding, and smaller farm animals.

Vasco da Gama   Lived 1460–1524. Portuguese navigator who in 1498 was the first to reach India by sea. He made a second voyage there in 1502. He was appointed viceroy of India in 1524 and died in Cochin.

venation   A mock hunt staged in the Roman amphitheater, in which men chased or fought with wild beasts.

Venice   An Italian city located in marshes at the head of the Adriatic Sea, which became a great center of trade and naval power in the Middle Ages. For much of that time it was ruled by *doges*, members of a merchant oligarchy. In modern times it has been threatened by pollution and flooding.

Villani, Giovanni   Lived 1275–1348. Florentine businessman and chronicler, he was author of the *Florentine History*, which provides information on the customs, industry, commerce, and arts of Florence. Died during the Black Plague.

Virgil (or Vergil), Publius Vergilius Maro   Lived 70–19 BC. A great Roman poet whose works include the *Aeneid* and the *Georgics*, the latter describing farm life and the round of the agricultural year.

water hyacinth   A weedy water plant, originally native to South America. Introduced to waters in many warmer parts of the world, probably because of its beautiful bluish flowers, it has formed extensive mats that choke harbors, reservoirs, and rivers, with negative impacts on fish and native plants.

water mill   A machine driven by a waterwheel and used to grind grain or for a variety of other industrial purposes requiring continuous power. Known from the time of the Roman Empire and used widely in the medieval Mediterranean.

Willcocks, William   Lived 1852–1932. A British engineer born and educated in India. He designed dams in India and the first Aswan Dam in Egypt, and prepared a water management scheme for Mesopotamia (Iraq).

windmill   A machine that directs wind energy to perform work. A prototype is known from the writings of Hero of Alexandria (AD 20–90), but the first widespread practical use in the Mediterranean came in the twelfth century.

wool   The most important commercial fiber for textiles in the Middle Ages. The demand for wool increased the numbers of sheep into the millions and resulted in great damage to the grasslands and soils of the Mediterranean basin, with resultant erosion.

world market economy   The system of international capital investment, monetary exchange, and free trade, with roots in the nineteenth century, which created agreements and institutions such as the International Monetary Fund, the World Bank, the General Agreement on Trade and Tariffs, and the World Trade Organization after World War II. Its purpose is to foster world financial and economic growth, so it has emphasized rapid development of natural resources rather than conservation and environmental values.

Zagros   A mountain range in Iran that separates the Mesopotamian lowlands from the Iranian plateau. Marks the eastern margin of Mesopotamia.

Zeus   Greek god of the sky and weather, including winds and lightning; father of gods and humans.

ziggurat   A structure resembling a step pyramid consisting of a series of superimposed platforms with a high temple at the top, accessible by stairways. Built of baked and unbaked bricks, it was provided with plantings of trees and other vegetation, and therefore fitted with drainage apertures.

CHRONOLOGY

BC

5 million The Mediterranean basin fills with water from the Atlantic Ocean through the opening Strait of Gibraltar.

300,000 This date represents the possible age of the Petralona skull, found in Greece, thought to be an intermediate form between *Homo erectus* and Neanderthal.

200,000 Hunters and gatherers range through the Mediterranean region.

100,000 Modern humans (*Homo sapiens*) enter the Mediterranean region.

28,000 Modern humans replace the Neanderthals throughout the Mediterranean region.

15,000 Humans first experiment with agriculture along the Nile, Tigris, and Euphrates rivers.

12,000 The Ice Age ends.

10,000 The Mediterranean climate approaches present conditions.

6000 The Agricultural Revolution spreads throughout the eastern Mediterranean.

5400 Wine is being made, preserved with resin, and kept in clay jars by this date in the Zagros Mountains of Iran.

4000–3000 The first cities in Mesopotamia and the Near East, including Egypt, are founded. Gilgamesh, king of Uruk, lives and rules sometime within this period.

3100 The unification of Egypt under one pharaoh and the beginning of the First Dynasty.

3000–1000 The Bronze Age.

2613–2494 The Old Kingdom in Egypt, age of the building of the great pyramids.

2300 Sargon the Great, king of Akkad, conquers Sumeria, founding the first Mesopotamian empire.

1700   The law code of Hammurabi, king of Babylon, regulates the use of water in agriculture.

1500   A map of Nippur, made in clay, shows a park in the center of the city.

1479–1425   Rule of Thutmose III, pharaoh of Egypt, who conquers a large part of the eastern Mediterranean coastlands. Apparently interested in natural history, he has images of many of the birds, animals, and plants of those areas carved on his temples.

1352–1336   Rule of Akhenaton, monotheistic pharaoh of Egypt. The "Amarna" art of his reign portrays animals and plants with naturalistic realism.

1000   The Iron Age begins (approximate).

530   An aqueduct tunnel designed by Eupalinus is built on the island of Samos, Greece.

429–347   Life of Plato, who leaves writings on many human interactions with the natural environment, including the deforestation of his native Attica.

396   The Carthaginian siege of Syracuse ends when a plague decimates the attacking army.

384–322   Life of Aristotle, who studies the natural phenomena of the Mediterranean region.

372–288   Life of Theophrastus, a Greek philosopher who writes on botany and ecology.

312   The first Roman aqueduct, the Aqua Appia, is built by the censor Appius Claudius.

AD

79   Pompeii and Herculaneum are buried by materials from the volcano Vesuvius.

164   The great plague begins during the reign of the Roman emperor Marcus Aurelius.

212   Caracalla enlarges the Roman citizen body and the tax rolls by an edict granting citizenship to all free men in the empire.

235–284   Period of the "military anarchy" in Rome, accompanied by environmental damage and ruinous inflation.

284–305   The emperor Diocletian reforms the Roman economy.

410   Sack of Rome by Alaric and the Visigoths.

476   Resignation of the emperor Romulus Augustulus; traditional date of the end of the western Roman Empire.

540   Bubonic plague sweeps the known world during the reign of the eastern Roman emperor Justinian I.

711   Muslims conquer Spain, later introducing new forms of irrigation agriculture.

750   The Abbasid caliphs make Baghdad their capital, inaugurating a flowering of literature, art, science, and medicine stimulated by a study of the Greek classics.

1050–1300   Europe experiences a period of rapid population growth and the expansion of agriculture.

1187   Saladin reconquers Jerusalem from the crusaders.

1204   The Fourth Crusade, at the instigation of Venice, captures Constantinople.

1250   Gunpowder arrives in the Mediterranean region (approximately).

1258   Mongols loot Baghdad, destroying its libraries.

1276   The Italian city of Verona passes a law to control water pollution.

1281   The city of Siena requires the planting of trees on private land.

1329 and 1339   Famine strikes the city of Florence.

1333   The Arno River floods due to deforestation in its watershed, causing devastation in Florence.

1347   The Black Death arrives in the Mediterranean region.

1425   Portuguese settlement begins on the island of Madeira.

1453   Constantinople falls to the Ottoman Turks, becoming the capital of the Ottoman Empire.

1464   Date of the first book to be printed in Italy.

1492   Christian Spain conquers Granada, completing the *Reconquista*.

1492   Christopher Columbus makes the first transatlantic voyage for Spain, beginning the "Columbian Exchange."

1498   Vasco da Gama, a Portuguese navigator, reaches India.

1500–1550   Maize (American corn) becomes a common crop in the Mediterranean area.

1526   Merino sheep under the Mesta, an association of owners, number 3.5 million in Spain.

1529   Ottoman Turks besiege Vienna, their furthest advance into Europe.

1530   Corks are first used in wine bottles.

1553   The potato is first mentioned by a European, Pedro Cieza de Leon, in Peru.

1560s   Coffee houses become popular in Istanbul.

1570s   English and Dutch ships enter the Mediterranean in significant numbers.

1609   The Moriscos, experts in irrigation agriculture, are expelled from Spain.

1632   The bark of the cinchona tree, from Peru, is recognized as effective against malaria.

1633   Galileo Galilei is imprisoned by the Roman Catholic Church for writing that the Earth revolves around the sun.

1669   The French Forest Ordinance applies conservation measures throughout France.

1692   The first cookbook to mention tomatoes is published in Naples.

1704   The British seize Gibraltar.

1730   The collapse of the "tulip craze," in which bulbs of the flowers bring hugely inflated prices.

1750   Beginning of the Industrial Revolution (approximately).

1768   France annexes Corsica.

1776   Edward Gibbon publishes *The Decline and Fall of the Roman Empire.*

1798   Napoleon enters Egypt. French scientists who travel with him study the environment of the country.

1805   Muhammad Ali seizes power in Egypt and begins state development of agriculture.

1821   Beginning of the Greek War of Independence against the Ottoman Empire.

1824   The French National School of Waters and Forests is founded in Nancy.

1831   Widespread cholera epidemic in the Mediterranean.

1858   Ottoman Land Law defines most agricultural land as state property.

1863   Phylloxera, a destructive aphid, appears in grapevines in France.

1864   George Perkins Marsh, an environmental historian, publishes *Man and Nature*, the first book to examine systematically the damaging changes produced by human impacts on the natural environment.

1869   The Suez Canal, designed by de Lesseps, is completed, joining the Mediterranean and Red seas.

1892   The last Iberian lynx in Portugal is shot and killed.

1902   The first Aswan dam on the Nile River is completed by the British.

1914–1918   World War I.

1918   The first Spanish national parks are created. Other Mediterranean nations soon follow by declaring their own national parks.

1920s   Exchange of populations between Greece and Turkey brings 1.6 million ethnic Greeks into Greece.

1927   Camargue Nature Reserve is created in southern France.

1938   Mount Olympus National Park is created in Greece.

1938–1939   Walter Clay Lowdermilk makes a survey of Mediterranean soil erosion and conservation, later publishing his findings as *Conquest of the Land through 7,000 Years.*

1939–1945   World War II.

1948   Israel is recognized by the United Nations.

1950s   Nature protection societies are organized in several Mediterranean nations. Chain saws become generally available in the developed nations.

1952 Gamal Abdel Nasser becomes president of Egypt as a result of revolution.

1960 Cyprus gains independence from the United Kingdom.

1960s Toxic "red tides" of algae become common in the Mediterranean Sea.

1960–1970 Construction of the Aswan High Dam in Egypt.

1962 Algeria gains independence from France.

1964 Malta becomes independent.

1966 The Arno River floods due to deforestation in its watershed, causing devastation in Florence. A high tide floods Venice disastrously.

1966 Fernand Braudel publishes, in French, *The Mediterranean and the Mediterranean World in the Age of Philip II.*

1970s Athens measures pollution levels higher than those of Los Angeles.

1971 Construction of the Aswan High Dam in Egypt is completed.

1973 The United Nations Environment Programme is established.

1975 The Mediterranean Action Plan is negotiated.

1976 The Barcelona Convention for the Protection of the Mediterranean Sea Against Pollution, or "Blue Plan," is signed by most Mediterranean nations.

1984 An underwater weed called *Caulerpa taxifolia* is first noticed in the Mediterranean Sea off Monaco.

1990s Thousands of striped dolphins die of disease in the western Mediterranean Sea.

1991 Beginning of the fragmentation of Yugoslavia.

1991 The European Community establishes the Mediterranean Desertification and Land Use (MEDALUS) program.

1991 The First Gulf War, in which a U.S.-led coalition forces an end to the Iraqi occupation of Kuwait. Iraqi military units deliberately cause oil spills and set fire to more than six hundred Kuwaiti oil wells, causing major environmental damage.

1993–2000 Drainage projects desiccate most of the marshes in the lower Tigris-Euphrates river system, destroying the way of life of the Marsh Arabs and the wetland ecosystems on which they depend.

1996 The International Convention on Combating Desertification is signed, including measures to study desertification in the Mediterranean region.

1997 The population of Cairo surpasses 10 million.

1997 In the Kyoto Protocol, major industrial nations agree to reduce their emissions of greenhouse gases by various amounts by 2012. The United States withdraws from the protocol in 2001.

2002 Pope John Paul II and Ecumenical Patriarch Bartholomew of Constantinople sign the Joint Declaration on Articulating a Code of Environmental Ethics in Venice, Italy, and Vatican City.

2002   The First Islamic Conference of Environmental Ministers is held in Jeddah, Arabia, and issues an Islamic Declaration on Sustainable Development, outlining a common view on several environmental issues.

2003   Start of the second Gulf War, in which a U.S.-led coalition defeats the Iraqi army and occupies Iraq. Environmental damage results from actions of both sides in the war.

2025   The expected population of the Mediterranean region for this year is 500 million, as compared with 150 million in 1950. Forests in the southern and eastern parts of the Mediterranean basin are expected to be gone by this date.

ANNOTATED BIBLIOGRAPHY

Abun-Nasr, J. M. 1971. *A History of the Maghrib.* Cambridge: Cambridge University Press.

> A standard history of Mediterranean North Africa, including Morocco, Algeria, and Tunisia.

Abu-Zaid, Mahmoud, and M. B. A. Saad. 1993. "The Aswan High Dam, 25 Years On." *UNESCO Courier* (May 1): 37.

> A reassessment of the Aswan High Dam and its economic and environmental effects.

Abu-Zeid, M., W. A. Charlie, D. K. Sunada, and A. Khafagy. 1995. "Seismicity Induced by Reservoirs: Aswan High Dam." *International Journal of Water Resources Development* 11, 2 (June): 205–213.

> A study of earthquakes caused by the weight of the water in Lake Nasser-Nubia upon the underlying strata and fault lines. The conclusion is that the earthquakes up to the date of publication had been moderate and had not caused major structural damage to the dam.

Agnoletti, Mauro. 2003. "Historical Research and Landscape Analysis: The Case of Tuscany, Italy." Prague: European Society for Environmental History, Second International Conference.

> Examination of changes in the landscape of Tuscany between 1832 and 2000, based on six study areas selected to represent the main geographical areas of the province. The investigators found deep changes in features, including the extent of woodlands and a decrease in landscape diversity.

Algaze, Guillermo. 1993. *The Uruk World System: The Dynamics of Expansion of Early Mesopotamian Civilization.* Chicago: University of Chicago Press.

> A regional study of the distribution and growth of settlements in southern Mesopotamia around the important Sumerian center of Uruk. The book

includes useful maps relating human settlements to natural and artificial watercourses.

Angel, J. Lawrence. 1972. "Ecology and Population in the Eastern Mediterranean." *World Archaeology* 4: 88–105.
    A key article establishing important parameters for the study of the relationship between human population and environmental resources.

Apostolopoulos, Yorghos, and Dennis J. Gayle, eds. 2002. *Island Tourism and Sustainable Development.* Westport, CT: Praeger.
    Includes a look at tourism on islands in the modern Mediterranean.

Apostolopoulos, Yorghos, Philippos Loukissas, and Lila Leontidou, eds. 2001. *Mediterranean Tourism: Facets of Socioeconomic Development and Cultural Change.* London: Routledge.
    Among the articles in this collection are some that address the effects of tourism on the Mediterranean environment.

Arsuaga, Juan Luis. 2001. *The Neanderthal's Necklace: In Search of the First Thinkers.* New York: Four Walls Eight Windows.
    A study based on Iberian archaeological evidence of the intelligence and technological capabilities of the Neanderthals.

Aruz, Joan, ed. 2003. *Art of the First Cities: The Third Millennium BC from the Mediterranean to the Indus.* New York: Metropolitan Museum of Art.
    The catalog for a major exhibit on the art of ancient Mesopotamia, the Levant, Iran, and India during the period of the growth and spread of urban civilization. This volume includes several useful introductory articles.

Ashby, Thomas. 1935. *The Aqueducts of Ancient Rome.* Oxford: Oxford University Press.
    A historical and geographical study of the construction and operation of the water supply system of the city of Rome in ancient times.

Attenborough, David. 1987. *The First Eden: The Mediterranean World and Man.* Boston and Toronto: Little, Brown.
    A popular but authoritative survey of the historical relationships between human societies and the natural world in the Mediterranean basin. This is the companion volume to a television program series with the same title.

Aubert, Jean-Jacques. 2001. "The Fourth Factor: Managing Non-Agricultural Production in the Roman World." In *Economies beyond Agriculture in the Classical World,* eds. David J. Mattingly and John Salmon, 90–112. London and New York: Routledge.

> Although the economy of the Roman Empire rested on an agricultural base, its structure included an important industrial component that is analyzed in this seminal article.

Baumann, Hellmut. 1993. *The Greek Plant World in Myth, Art, and Literature.* Portland, OR: Timber Press.

> Filled with anecdotes and well illustrated, this is a fascinating excursion into the significance of numerous species of plants in Greek mythology, letters, and the arts.

Baynes, N. H. 1943. "Decline of the Roman Empire in Western Europe: Some Modern Explanations." *Journal of Roman Studies* 33: 29–35.

> An enlightening exercise in "cliometrics," the use of quantifiable changes in history. Baynes uses statistics derived from everything from inscriptions and literature to underwater archaeology.

Bechmann, Roland. 1990. *Trees and Man: The Forest in the Middle Ages.* New York: Paragon House.

> A definitive and interestingly written study that includes many references to trees, forestry, and forest products in the Mediterranean region. This is the indispensable source on medieval forest history.

Blondel, Jacques, and James Aronson. 1999. *Biology and Wildlife of the Mediterranean Region.* Oxford: Oxford University Press.

> A broad-ranging study, accessible to scholars who are not professionals in the field of biology, on biodiversity, habitats, and populations. The authors include a useful chapter on the history of humans as sculptors of the Mediterranean landscape.

Boak, Arthur E. R. 1955. *Manpower Shortage and the Fall of the Roman Empire in the West.* Ann Arbor: University of Michigan Press.

> A classic study of the relationship between population and economics.

Bolle, Hans-Jürgen, ed. 2003. *Mediterranean Climate: Variability and Trends.* Berlin: Springer-Verlag.

A collection of articles, many of them informative, on the parameters and history of the Mediterranean climatic zone.

Booth, Martin. 1996. *Opium: A History.* New York: St. Martin's Press.
Traces the long story of the many human uses, some of them very ancient, of the chemical products of the poppy plant, *Papaver somniferum.*

Borowiec, Andrew. 2003. *Taming the Sahara: Tunisia Shows a Way while Others Falter.* Westport, CT: Praeger.
Chronicles a number of contemporary projects for the reclamation of desert lands in Tunisia.

Bowlus, Charles R. 1980. "Ecological Crises in Fourteenth-Century Europe." In *Historical Ecology: Essays on Environment and Social Change,* ed. Lester J. Bilsky, 86–99. Port Washington, NY: Kennikat Press.
An article based on a wide variety of medieval sources that makes the case for deforestation due to unsustainable agricultural expansion, causing a series of economic and social emergencies.

Brandt, C. Jane, and John B. Thornes, eds. 1996. *Mediterranean Desertification and Land Use.* Chichester, UK: John Wiley and Sons.
These articles explain the concerns and discoveries that motivated and resulted from the Mediterranean Desertification and Land Use programs of the European Community.

Braudel, Fernand. 1972. *The Mediterranean and the Mediterranean World in the Age of Philip II.* New York: Harper and Row.
This monumental work has been widely hailed as one of the first and most influential works to argue the importance of environmental factors in the long-term history of the Mediterranean basin. Its emphasis, however, is more on continuity than on change.

Bulliet, Richard W. 1990. *The Camel and the Wheel.* New York: Columbia University Press.
The story of the domestication and use of the camel for transportation throughout North Africa and the Near East.

Butzer, Karl W. 1990. "The Realm of Cultural-Human Ecology: Adaptation and Change in Historical Perspective." In *The Earth as Transformed by Human Action,* ed. B. L. Turner, 685–701. New York: Cambridge University Press.

A theoretical overview of human-environmental interactions, based on archaeological and historical knowledge, covering chiefly the eighteenth, nineteenth, and twentieth centuries.

Camuffo, Dario, Emanuela Pagan, Giovanni Sturaro, and Giovanni Cecconi. 2003. "Change in Sea-Level and Storm Surge Frequency at Venice from Proxy: The Problem and the Impact on the Historical Buildings." In *Dealing with Diversity: Proceedings of the 2nd International Conference of the European Society for Environmental History*, eds. by Leoš Jelecek et al., 29–32. Prague: Charles University, Faculty of Science.

An assessment of the experience of Venice from tidal flooding surges using instrumental observations over the period 1872–2000 and documentary data over the past millennium. Photographic evidence is also used to establish that the damage due to surges and rising mean sea level is unsustainable.

Carrington, Richard. 1971. *The Mediterranean: Cradle of Western Culture*. New York: Viking Press.

A well-illustrated popular and informative survey of various aspects of Mediterranean human history and natural history.

Carroll, John E., ed. 1988. *International Environmental Diplomacy: The Management and Resolution of Transfrontier Environmental Problems*. Cambridge: Cambridge University Press.

Discussed among the issues transcending national borders are the treatment of the Tigris-Euphrates dams, diversions, and the allocation of water resources.

Charlesworth, Martin P. 1951. "Roman Trade with India: A Resurvey." In *Studies in Roman Economic and Social History*, ed. P. R. Coleman-Norton, 131–143. Princeton: Princeton University Press.

An important causative factor often alleged for the decline of the Roman imperial economy is the drain of precious metals to the east in return for mainly luxury goods. This article provides evidence to support that assumption.

Chew, Sing C. 2001. *World Ecological Degradation*. Lanham, MD: Rowman and Littlefield.

This sociologically based analysis of world environmental history from the earliest times emphasizes deforestation, urbanization, and the unequal distribution of resources as factors resulting from and modifying human economic and political activities.

Cipolla, Carlo M. 1982. *The Monetary Policy of Fourteenth-Century Florence.* Berkeley and Los Angeles: University of California Press.
  A study indispensable to the understanding of the interaction between the central role of Florence in the medieval European economy and the consumption of resources.

Collins, Robert O. 1990. *The Waters of the Nile: Hydropolitics and the Jonglei Canal, 1900–1988.* Oxford: Clarendon Press.
  When completed, the Jonglei canal will reorganize a large section of the hydrology of the upper Nile, with far-reaching effects on the ecology and biodiversity of the Nile basin. This study explains many of the political aspects of the decision to build the project and the reasons for its delay.

Concina, Ennio. 1989. "Humanism on the Sea." *Mediterranean Historical Review* 3, 1 (Special Issue: Mediterranean Cities: Historical Perspectives): 159–165.
  A reflection on the setting of humanistic efforts in the Mediterranean environment.

Corcoran, Simon 1996. *The Empire of the Tetrarchs.* Oxford: Clarendon Press.
  Surveys the political and economic world of the Roman Mediterranean in the third and fourth centuries AD.

Crosby, Alfred W., Jr. 1972. *The Columbian Exchange: Biological and Cultural Consequences of 1492.* Westport, CT: Greenwood Press.
  This is one of the most influential books written on environmental history; it surveys the impact of contact between the Old and New Worlds in terms of the transfer of animals, plants, and microorganisms in both directions, although most importantly from east to west.

————. 1984. "An Ecohistory of the Canary Islands." *Environmental Review* 8, 3 (Fall): 214–235.
  The Canaries, islands with a Mediterranean climate, were among the first places to be colonized by Europeans and altered in terms of biology, including human and nonhuman inhabitants. Thus they offer Crosby an early example of his thesis of ecological imperialism.

Crouch, Dora P. 1993. *Water Management in Ancient Greek Cities.* New York: Oxford University Press.

> Crouch emphasizes the importance of limestone stratigraphy and karst landscapes in the attempts of Greek cities to exploit watersheds and secure urban water supplies.

Crouzet, François. 2001. *A History of the European Economy, 1000–2000.* Charlottesville: University Press of Virginia.

> A survey including case studies of crucial epochs in the development of the agricultural and nonagricultural economy in Europe.

Damianos, Dimitris, Efthalia Dimara, Katharina Hassapoyannes, and Dimitris Skuras. 1998. *Greek Agriculture in a Changing International Environment.* Aldershot, UK: Ashgate.

> This book is a study of the modern Greek economy and includes analysis of the issues involved in integration into the European Community.

Davisson, William I., and James E. Harper. 1972. *European Economic History. Volume I: The Ancient World.* New York: Appleton Century Crofts.

> Beginning with the Neolithic period, this work surveys the early river valley civilizations, the Near East, Egypt, Greece (including the economic thought of Aristotle), Alexander and the Hellenistic world, and Rome. The authors give some attention to technological innovations.

Diamond, Jared. 1997. *Guns, Germs, and Steel: The Fates of Human Societies.* New York: Norton.

> A compelling argument, worldwide in scope and amazing in detail, for environmental determinism based on the availability of resources and domesticable animals. Diamond discounts the roles of race and differential intelligence in the success of some human groups at the expense of others.

Drachmann, Aage Gerhardt. 1948. *Ktesibios, Philon, and Heron.* Copenhagen: E. Munksgaard.

> A study of technological invention in ancient Alexandria, emphasizing pneumatics.

———. 1963. *The Mechanical Technology of Greek and Roman Antiquity: A Study of the Literary Sources.* Madison: University of Wisconsin Press.

> A good reference work for those studying ancient technologies.

Dregne, H. E. 1983. *Desertification of Arid Lands.* Chur, Switzerland (Confederatio Helvetica): Harwood Academic Publishers.

>Dregne argues that for many parts of the world, including North Africa, desertification usually means the advance of the desert boundary, but in Europe it involves desiccation of smaller areas and changes in vegetation.

Duncan-Jones, Richard P. 1982. *The Economy of the Roman Empire: Quantitative Studies.* Cambridge: Cambridge University Press.

>An interesting exercise in so-called cliometrics, the use of quantifiable data in historical studies.

———. 1990. *Structure and Scale in the Roman Economy.* Cambridge: Cambridge University Press.

>Useful in explaining the relationship between Roman social structure and the organization of the economy.

Elver, Hilal. 2000. *Peaceful Uses of International Rivers: The Euphrates and Tigris Rivers Dispute.* Ardsley, NY: Transnational Publishers.

>A comprehensive study of water development issues between Iraq and its neighbors before the structure of international agreements was dismantled by the second Gulf War.

Evans, J. K. 1981. "Wheat Production and Its Social Consequences in the Roman World." *Classical Quarterly* 31: 428–442.

>An analysis of ancient Rome's dependence on grain importation from other areas within Roman dominions to feed the underclasses of the city and its environs.

Fahim, Hussein M. 1981. *Dams, People, and Development: The Aswan High Dam Case.* New York: Pergamon Press.

>An incisive look at the way in which the decision to build the Aswan High Dam was dominated by political rather than engineering and environmental considerations.

Faulkner, Hazel, and Alan Hill. 1997. "Forests, Soils, and the Threat of Desertification." In *The Mediterranean: Environment and Society*, eds. Russell King, Lindsay Proudfoot, and Bernard Smith, 252–272. London: Arnold.

>This important article explores the degradation of the Mediterranean environment by human action, including erosion and desertification due to changing patterns of land use and the effects of global warming.

Finley, M. I. 1999. *The Ancient Economy*. Berkeley and Los Angeles: University of California Press.

> A revised and updated view on the economic development of the ancient world.

Franghiadis, Alexis. 2003. "Commons and Change: The Case of the Greek 'National Estates'" (nineteenth-early twentieth centuries). Prague: European Society for Environmental History, Second International Conference.

> With independence from the Ottoman Empire, the new Greek government found itself in possession of public agricultural lands that had formerly belonged to Ottoman landlords. They were treated as common lands accessible to peasants in an unrestricted manner, with surprisingly positive results, according to the author, showing that the reinforcement of private property is not always the only solution to economic problems.

Friedländer, L. 1909. *Roman Life and Manners under the Early Empire*. 4 vols. London: George Routledge.

> An exhaustive study of sources then available for the study of Roman social history, although it must be supplemented by reference to more recent archaeological studies.

Geeson, N. A., C. J. Brandt, and J. B. Thornes, eds. 2002. *Mediterranean Desertification: A Mosaic of Processes and Responses*. Chichester, UK: John Wiley and Sons.

> "Mosaic" in the title of this book may be taken as referring to the diverse character of the Mediterranean landscape and to the complex set of factors involved in the present crisis of desertification. Each of these meanings is well investigated by the authors.

Gimpel, Jean. 1976. "Environment and Pollution." In *The Medieval Machine: The Industrial Revolution of the Middle Ages*, 75–81. New York: Holt, Rinehart, and Winston.

> Pollution is often thought to be mainly a product of technological development in the modern world, but this succinct chapter abundantly demonstrates that thought to be a misconception.

Glacken, Clarence. 1967. *Traces on the Rhodian Shore*. Berkeley and Los Angeles: University of California Press.

> Deftly combining intellectual history with environmental philosophy, this pioneering study investigates the history of three ideas—the Earth as an

environment designed for humans, environmental influences on humans, and human causation of environmental changes—in the West from ancient Greek and Roman times through the eighteenth century.

Goldsmith, Edward, and Nicholas Hildyard. 1984. *The Social and Environmental Effects of Large Dams.* San Francisco: Sierra Club Books.
> A critique of the obsession of twentieth-century nation-states with the building of ever larger dams for power generation and irrigation, but also for political aggrandizement. One of the most notable of these modern pyramids was, of course, the Aswan High Dam in Egypt.

Gould, Stephen Jay. 2003. "Syphilis and the Shepherd of Atlantis." In *I Have Landed: The End of a Beginning in Natural History*, 192–212. New York: Three Rivers Press.
> An examination of the widely accepted idea that syphilis was introduced into Europe by Spanish sailors from the Columbian expeditions of discovery and exploration in the Americas.

Grenon, Michel, and Michel Batisse, eds. 1989. *Futures for the Mediterranean Basin: The Blue Plan.* Mediterranean Action Plan and United Nations Environment Programme. Oxford: Oxford University Press.
> A collection of articles on the plan organized by the United Nations Environment Programme for the environmental protection and restoration of the Mediterranean area, which gained the cooperation of virtually every one of the often mutually hostile states with Mediterranean seacoasts.

Grmek, Mirko D. 1989. *Diseases in the Ancient Greek World.* Baltimore: Johns Hopkins University Press.
> A study of diseases described in ancient Greek texts, including the noted plague of Athens described in clinical detail by Thucydides in his *Peloponnesian War*.

Haas, Peter M. 1990. *Saving the Mediterranean: The Politics of International Environmental Cooperation.* New York: Columbia University Press.
> This careful study indicates that it is easier to get nation-states to agree to far-reaching programs of international cooperation than to fund them and enforce them.

Hall, Marcus. 1998. "Restoring the Countryside: George Perkins Marsh and the Italian Land Ethic (1861–1882)." *Environment and History* 4, 1 (February): 91–104.

> Hall maintains that Marsh, considered by many scholars to be the founder of American environmental history, spent much of his life in Italy, was influenced by Italian ideas, and in turn had influence upon Italian land managers during the nineteenth century.

Hastings, Tom H. 2000. *Ecology of War and Peace.* Lanham, MD: University Press of America.

> An important study of the effects of war on landscape and the environment.

Hauck, George F. W. 1989. "The Roman Aqueduct of Nimes." *Scientific American* 260 (March): 98–104.

> This well-explained article describes the construction and function of one of the most notable surviving Roman aqueducts, including the careful calculation of the grade of descent and the provision of alternate facilities allowing the cleaning of the system and the removal of silt without interrupting the delivery of water to public and private facilities.

Hehn, Victor. 1976. *Cultivated Plants and Domestic Animals in Their Migration from Asia to Europe: Historico-Linguistic Studies.* Amsterdam: John Benjamins B.V.

> A study of the introduction of Asiatic domestic species into European agriculture, based on linguistic and historical evidence.

Helfand, Jonathan. 1986. "The Earth Is the Lord's: Judaism and Environmental Ethics." In *Religion and Environmental Crisis,* ed. Eugene C. Hargrove, 38–52. Athens: University of Georgia Press.

> Helfand uses both scripture and tradition to elucidate Jewish environmental ethics, including prominently the principle of *bal tashhit,* "you shall not despoil," and the use of the principles of environmental protection in the settlement of the land.

Helms, J. Douglas. 1984. "Walter Lowdermilk's Journey: Forester to Land Conservationist." *Environmental Review* 8, 2 (Summer): 132–145.

> Lowdermilk, a founder of the U.S. Soil Conservation Service, served as an advisor in China and made a survey of agriculture and soils in lands around the Mediterranean in the years just before World War II. After the war, he

provided technical assistance to soil conservationists in Israel. This article evaluates his achievements.

Herlihy, David. 1997. *The Black Death and the Transformation of the West.* Cambridge, MA: Harvard University Press.
> The social and environmental effects of the Black Death are described in this well-researched book.

Hillel, Daniel J. 1991. *Out of the Earth: Civilization and the Life of the Soil.* New York: Free Press.
> A study of the history of soil erosion and soil conservation, including important observations on the ancient world and the Mediterranean region.

———. 1994. *Rivers of Eden: The Struggle for Water and the Quest for Peace in the Middle East.* New York: Oxford University Press.
> Indicates the critical relationship between water development and allocation and peace in the Jordan and Tigris-Euphrates watersheds.

Hoffmann, Richard C. 1995. "Environmental Change and the Culture of Common Carp in Medieval Europe." *Guelph Ichthyology Reviews* 3 (May): 57–85.
> Hoffmann, the leading authority on fish culture and fishing in medieval Europe, describes the propagation of an important food source and its environmental relationships.

Hopkins, Keith. 1988. "Roman Trade, Industry, and Labor." In *Civilization of the Ancient Mediterranean: Greece and Rome,* vol. 2, eds. Michael Grant and Rachel Kitzinger, 753–778. New York: Charles Scribner's Sons.
> A comprehensive survey of population, labor (including the role of slavery), technology, mining, transportation, and trade by land and sea in the Roman Empire.

Hsu, Kenneth J. 1983. *The Mediterranean Was a Desert.* Princeton: Princeton University Press.
> A discussion of the geological evidence indicating that the Mediterranean almost completely evaporated when the Strait of Gibraltar closed due to tectonic movement at one or several times during the Cenozoic era.

Hughes, J. Donald. 1975. *Ecology in Ancient Civilizations.* Albuquerque: University of New Mexico Press.

A survey of human-environment relationships in the ancient world, including Egypt and the Near East, Judaism and Christianity, and Greece and Rome.

———. 1976. "The Effect of Classical Cities on the Mediterranean Landscape." *Ekistics* 42: 332–342.

The environmental aspects of city planning, urban activities, agriculture, technology, and exploitation directed by the needs of Greek and Roman cities are investigated in this article.

———. 1988. "Land and Sea." In *Civilization of the Ancient Mediterranean: Greece and Rome*, vol. 1, eds. Michael Grant and Rachel Kitzinger, 89–133. New York: Charles Scribner's Sons.

This outline of the ancient Mediterranean environment includes climate, topography and geology, the sea, vegetation, animal life, and the effects of human activities on all of these.

———. 1992. "Sustainable Agriculture in Ancient Egypt." *Agricultural History* 66 (Spring): 12–22.

This article maintains that Egyptian civilization was generally in harmony with environmental factors that encouraged it, although changes were sometimes damaging and crises occurred.

———. 1994. *Pan's Travail: Environmental Problems of the Ancient Greeks and Romans*. Baltimore and London: Johns Hopkins University Press.

A study of impacts on the environment in classical antiquity and reciprocal effects on the society of the Greeks and Romans, including deforestation, wildlife depletion, industrial pollution, agricultural decline, and urban problems. The role of environmental problems as factors in the decline of ancient civilizations is discussed.

———. 1996. "Francis of Assisi and the Diversity of Creation." *Environmental Ethics* 18, 3 (Fall): 311–320.

Francis is a definitive figure in the delineation of medieval attitudes toward the environment, embodying the positive elements of the affirmation of creation, and respect for the diverse forms of natural creatures.

———. 2001. *An Environmental History of the World: Humankind's Changing Role in the Community of Life*. London and New York: Routledge.

A survey of global environmental history, this book includes case studies of Mesopotamia, Egypt, Athens, and Rome.

————. 2003. "Europe as Consumer of Exotic Biodiversity: Greek and Roman Times." *Landscape Research* (UK) 28, 1 (January): 21–31.

> Introductions of animals from Africa and Asia to the European section of the ancient Mediterranean through trade, adoption for domestic purposes, exhibition, and exploitation in the Roman amphitheatre are discussed in this article.

Hunt, Edwin S. 1994. *The Medieval Super-Companies.* Cambridge: Cambridge University Press.

> The forerunners of capitalism, including the Florentine houses of Bardi and Peruzzi, are the subjects of this study of the transformation of the medieval economy.

Hurst, H. E. 1957. *The Nile: A General Account of the River and the Utilization of Its Waters.* London: Constable.

> A valuable review of the history of the control of the Nile from the time of Muhammad Ali to Gamal Abdel Nasser.

Inälcik, H., and D. Quataert. 1994. *An Economic and Social History of the Ottoman Empire, 1300–1914.* Cambridge: Cambridge University Press.

> Although the Ottoman Empire is often regarded as an example of resistance to modernization, this study indicates that there were important changes during its long history.

Innocent III (Lotharia dei Segni). 1969. *On the Misery of the Human Condition.* Ed. Donald R. Howard. Indianapolis: Bobbs-Merrill.

> The medieval papacy reached the apex of its secular power under Innocent III, who favored the efforts of St. Francis, but this document indicates that his attitude toward the natural world contrasted significantly with that of the saint.

Jacobsen, Thorkild, and Robert M. Adams. 1958. "Salt and Silt in Ancient Mesopotamian Agriculture." *Science* 128: 1251–1258.

> An examination of the evidence for salinization and erosional deposition as factors in the decline of fertility in ancient Mesopotamia.

Jefitic, L., J. D. Milliman, and G. Sestini, eds. 1992. *Climatic Change and the Mediterranean.* London: Edward Arnold.

A collection of articles that are studies of the effects of changes in the regimes of temperature and precipitation on the agriculture of the Mediterranean area.

Jennison, G. 1937. *Animals for Show and Pleasure in Ancient Rome.* Manchester, UK: Manchester University Press.

> A study of the use of animals in the Roman amphitheatre. "Show and pleasure" is a euphemism for the cruelty, sadism, and death that were entailed in the Roman industry of the arena.

Johnson, William M. 2004. *Monk Seals in Post-Classical History.* Leiden: Nederlandshe Commissie voor Internationale Natuurbescherming, Mededelingen No. 39.

> The only survey in existence of references in medieval and modern times concerning the highly endangered monk seal.

Johnson, William M., and David M. Lavigne. 1999. *Monk Seals in Antiquity.* Leiden: Nederlandshe Commissie voor Internationale Natuurbescherming, Mededelingen No. 35.

> Johnson and Lavigne have assembled virtually every ancient reference to the monk seal, an important marine mammal formerly distributed throughout the Mediterranean.

Jordan, William Chester. 1996. *The Great Famine: Europe in the Early Fourteenth Century.* Princeton: Princeton University Press.

> An authoritative investigation, with excellent statistical support of the causes, extent, and results of the famines that swept Europe in the 1300s.

Keahey, John. 2002. *Venice against the Sea: A City Besieged.* New York: St. Martin's Press.

> An account of the various surveys and projects to protect Venice against the threats of storm surges and rising sea level in the twentieth century.

Khalid, Fazlun M., and Joanne O'Brien, eds. 1992. *Islam and Ecology.* London: Cassell.

> Articles in this elegant selection include expositions of environmental ethics, science, natural resources, trade, and conservation in Islamic thought and practice.

King, Russell. 1990. "The Mediterranean: An Environment at Risk."
Geographical Viewpoint 18: 5–31.
>   A seminal article identifying the salient problems of the modern Mediterranean.

King, Russell, Paolo de Mas, and Jan Mansvelt Beck, eds. 2001. *Geography, Environment, and Development in the Mediterranean.* Brighton, UK: Sussex Academic Press.
>   A remarkably comprehensive collection identifying the most important predicaments of environment and development in the modern Mediterranean lands.

King, Russell, Lindsay Proudfoot, and Bernard Smith, eds. 1997. *The Mediterranean: Environment and Society.* London: Arnold.
>   One of the most authoritative and indispensable books on the entire sweep of Mediterranean environmental history. This collection includes articles on every period of Mediterranean development from early to contemporary times.

Kliot, Nurit. 1994. *Water Resources and Conflict in the Middle East.* London: Routledge.
>   Traces the decisive yet often neglected relationships between aridity and scarce water with interstate rivalries and warfare.

Krech, Shepard, III, J. R. McNeill, and Carolyn Merchant, eds. 2004. *Encyclopedia of World Environmental History.* New York: Routledge.
>   A very important reference work containing articles on many subjects and issues relevant to the study of Mediterranean environmental history.

Lamb, H. H. 1995. "Through Viking Times to the High Middle Ages" and "Decline Again in the Late Middle Ages." In *Climate, History, and the Modern World*, 2nd ed., 171–210. London: Routledge.
>   Lamb is the authority on climatic change in Europe and globally in every period. These studies elucidate the period of the "medieval optimum" and the disastrous times that succeeded.

Lanoix, J. N. 1958. "Relation between Irrigation Engineering and Bilharziasis." *Bulletin of the World Health Organization* 18.

Bilharziasis (schistosomiasis) is a disease whose spread is attributed to the changes in water regime due to the construction of the Aswan High Dam in Egypt. This study shows some of the reasons for that conclusion.

Lees, G. M., and N. L. Falcon. 1952. "The Geographical History of the Mesopotamian Plains." *The Geographical Journal* 118 (March): 24–39.
This study provides geological evidence against the idea that the coastline of the upper Persian Gulf has moved considerably in a seaward direction as a result of the deposition of erosional sediments. Other studies contradict its conclusions.

Lewis, Bernard. 1982. *The Muslim Discovery of Europe.* New York: W. W. Norton and Company.
Analyzes the encounter between Islam and the West from the Islamic point of view. In this study, Islam is the active partner in the confrontation.

Llewellyn, Othman Abd ar-Rahman. 1984. "Islamic Jurisprudence and Environmental Planning." *Journal of Research in Islamic Economics* 1, 2 (Winter): 25–49.
Places environmental issues within the context of Islamic legal tradition.

Long, Pamela O. 2003. *Technology and Society in the Medieval Centuries: Byzantium, Islam, and the West, 500–1300.* Washington, DC: American Historical Association.
A historiographical guide to technological development in the medieval period throughout the Mediterranean area.

Lowdermilk, Walter C. 1953. *Conquest of the Land through 7,000 Years.* U.S. Department of Agriculture, Soil Conservation Service, Agriculture Information Bulletin No. 99.
The results of a field survey undertaken in the years just preceding World War II by one of the founders of the science of soil conservation in the United States. Lowdermilk concluded that the decline of forests and fertility in the Mediterranean had begun in ancient times and still continues in the contemporary milieu.

Lowenthal, David. 2000. *George Perkins Marsh: Prophet of Conservation.* Seattle: University of Washington Press.
The definitive work on the diplomat and writer who is often considered to be the morning star of environmental history.

Macmullen, Ramsay. 1988. *Corruption and the Decline of Rome.* New Haven, CT: Yale University Press.

> One of the causes often advanced for the decline and fall of the Roman Empire is moral corruption. Macmullen investigates this idea and offers a more comprehensive and analytical view.

Mairota, P., J. B. Thornes, and N. Geeson, eds. 1998. *Atlas of Mediterranean Environments in Europe: The Desertification Context.* Chichester, UK: John Wiley.

> A convincing assemblage of studies indicating that desertification is an important ongoing process in the contemporary Mediterranean basin.

Mann, Michael E., Raymond S. Bradley, and Malcolm K. Hughes. 1998. "Global-Scale Temperature Patterns and Climate Forcing over the Past Six Centuries." *Nature* 392 (23 April): 779–787.

> A contribution to the ongoing discussion concerning the relative importance of natural climatic change and human activities in the historical climatic regime.

Marsh, George Perkins. 1864, reprinted 1965. *Man and Nature.* Ed. David Lowenthal. Cambridge, MA: Harvard University Press.

> The definitive edition of the first great statement of the principle of environmental history: that human actions cause changes in the natural environment and that most of these changes are deleterious. According to Marsh, "Man is the disturber of nature's harmonies."

Mattingly, David J., and John Salmon, eds. 2001. *Economies beyond Agriculture in the Classical World.* London: Routledge.

> A look at manufacturing and trade in the Greek and Roman worlds.

McCormick, Michael. 2001. *Origins of the European Economy: Communications and Commerce, AD 300–900.* Cambridge: Cambridge University Press.

> An excellent recent study of a critical period in the history of the European economy, including detailed consideration of Mediterranean southern Europe.

McGovern, Patrick. 2003. *Ancient Wine: The Search for the Origins of Viniculture.* Princeton: Princeton University Press.

McGovern examines the archaeology of winemaking and finds its place of origin in western Asia.

McNeill, John R. 1992. *The Mountains of the Mediterranean World: An Environmental History.* Cambridge: Cambridge University Press.

An examination of historical degradation of mountain environments in the Mediterranean region, using five areas as case studies: the Taurus Mountains of southern Turkey; the Pindos Range in northwestern Greece; the Lucanian Apennines of southern Italy; the Sierra Nevada in Andalucia, Spain; and the Rif Mountains or Little Atlas of northern Morocco. McNeill argues that depopulation of these regions is in part a response to environmental decline.

———. 2000. *Something New under the Sun: An Environmental History of the Twentieth-Century World.* New York: W. W. Norton and Company.

A tour de force in which McNeill argues that the twentieth century is fundamentally different from all previous centuries in that the human species has inadvertently undertaken a gigantic uncontrolled experiment on the world environment. Chapters discuss each of the Earth's major "spheres" and the three great engines of change (urbanization/population, energy/economics, and ideas/politics).

McNeill, William H. 1976. *Plagues and Peoples.* Garden City, NY: Doubleday.

A worldwide study of the effects of communicable disease on human history.

Meiggs, Russell. 1982. *Trees and Timber in the Ancient World.* Oxford: Clarendon Press.

A masterpiece of detective work in scattered and little-known sources for the history of forestry, the use of forest products, and the timber trade in the ancient Mediterranean world.

Merlin, Mark David. 1984. *On the Trail of the Ancient Opium Poppy.* London: Associated University Presses.

A study of the cultivation and uses of one of the earliest narcotic plants known to humankind.

Morgan, David O. 1990. "The Mongols and the Eastern Mediterranean." *Mediterranean Historical Review* 4, 1: 198–211.

A study of the "second wave" Mongol invasions of the Levant in the thirteenth century.

Morley, Neville. 1996. *Metropolis and Hinterland: The City of Rome and the Italian Economy, 200 BC–AD 200.* Cambridge: Cambridge University Press.

> An economic history of the late Roman Republican and early Imperial periods, emphasizing the relationships between urban centers and the countryside.

Mundy, John H. 1973. "Food Production and Famine." In *Europe in the High Middle Ages, 1150–1309,* 115–116. New York: Basic Books.

> A brief but keen observation on the pattern of population increase, land use, and food supply in the period leading up to the Great Famine.

Munn, Ted, ed. 2002. *Encyclopedia of Global Environmental Change.* 5 vols. Hoboken, NJ: Wiley.

> The five volumes of this multidisciplinary reference work focus in turn on physical and chemical dimensions, biology and ecology, causes and consequences, response to global environmental change, and social and economic dimensions.

Murakami, Masahiro. 1995. *Managing Water for Peace in the Middle East: Alternative Strategies.* Tokyo: United Nations University Press.

> This book features a plan for the joint development of the Jordan River, the Dead Sea, and the Aqaba region, detailing alternatives for the transboundary transport of water. It highlights cogeneration applications and the political, economic, and technical viability of the strategic use of such sources as brackish water, seawater, and reclaimed waste water.

Nasr, Seyyed Hossein. 1968. *The Encounter of Man and Nature: The Spiritual Crisis of Modern Man.* London: George Allen and Unwin.

> A study of the history and philosophy of nature and science in various religious traditions from the standpoint of an eminent and articulate Islamic scholar.

———. 1973. "The Ecological Problem in the Light of Sufism: The Conquest of Nature and the Teachings of Eastern Science." In *Sufi Essays,* 152–163. Albany: State University of New York Press.

> The Islamic tradition of Sufism is presented as a key to the problem of a theology of nature in the context of comparative religion, both East and West.

Nicholson, Emma, and Peter Clark. 2003. *The Iraqi Marshlands: A Human and Environmental Study*. London: Politico's Publishing.
The marshlands of southern Iraq, an area of historical and ethnographic importance from early times but a disaster area in environmental and human terms from the 1990s onward, are clearly depicted in this excellent short book.

Organisation for Economic Co-operation and Development. 1983. *Environmental Policies in Greece*. Paris.
An official report that presents a reasonably balanced picture of governmental environmental policies in Greece in the 1980s.

Palmer, William G. 1984. "Environment in Utopia: History, Climate, and Time in Renaissance Environmental Thought." *Environmental Review* 8, 2 (Summer): 162–178.
Palmer's study of environmental philosophy in the Renaissance concentrates on the writings of Machiavelli, Jean Bodin, Thomas More, Francis Bacon, and Gerrard Winstanley.

Pamuk, Shevket. 1987. *The Ottoman Empire and European Capitalism, 1820–1913*. Cambridge: Cambridge University Press.
Traces the disastrous course of the Ottoman economy in its confrontation with western Europe in the period leading up to World War I.

Pavé, Marc. 2003. "History of the Sustainable Management of the South-West European Fishing." In *Dealing with Diversity: Proceedings of the 2nd International Conference of the European Society for Environmental History*, eds. Leoš Jelecek et al., 87–90. Prague: Charles University, Faculty of Science.
Traces the development of a persistent organization controlling fishing in the French Mediterranean Sea between 1715 and 1850, the *prud'homies* or fishermen's labor relations boards. This model persists today around the Mediterranean.

Perlin, John. 1989. *A Forest Journey: The Role of Wood in the Development of Civilization*. New York: W. W. Norton.
Perlin traces the history of wood use, supply, and deforestation in the development of Western civilization, emphasizing important parts of the Near East, Mediterranean, Europe, and the United States.

Pinkele, Carl F., and Adamantia Polis. 1983. *The Contemporary Mediterranean World.* New York: Praeger Publishers.
> Among the many articles in this substantial collection are useful studies of the environment, labor, economy, development, and the international context.

Pirenne, Henri. 1939. *Mohammed and Charlemagne.* London: Allen and Unwin.
> The classic statement of the Pirenne thesis that the economy of the Roman Mediterranean did not fundamentally change until after the Arabic conquests.

Polunin, Oleg, and Anthony Huxley. 1966. *Flowers of the Mediterranean.* Boston: Houghton Mifflin.
> A complete and authoritative guidebook not only to flowers but to higher plants including trees.

Postan, M. M. 1973. *Medieval Trade and Finance.* Cambridge: Cambridge University Press.
> An influential work on the social and economic history of the Middle Ages. This collection represents Postan's major contributions. Twenty-two essays are gathered in two volumes: *Essays on Medieval Agriculture and General Problems of the Medieval Economy and Medieval Trade and Finance.*

Prager, Frank D. 1978. "Vitruvius and the Elevated Aqueducts." *History of Technology* 3: 105–121.
> Vitruvius, an ancient Latin writer, gives an architectural description of the construction of the Roman water supply and distribution systems.

Pritchard, James B., ed. 1958. *The Ancient Near East: An Anthology of Texts and Pictures.* Princeton: Princeton University Press.
> A still useful collection of short illustrative texts from Mesopotamia, Egypt, and the Levant, many illustrating environmental attitudes.

Proudfoot, Lindsay. 1997. "The Graeco-Roman Mediterranean." In *The Mediterranean: Environment and Society,* eds. Russell King, Lindsay Proudfoot, and Bernard Smith, 57–74. London: Arnold.
> A survey of settlements in the Greek, Hellenistic, and Roman periods (800 BC–AD 395), including remarks on the end of the Roman Empire.

Proudfoot, Lindsay, and Bernard Smith. 1997. "From the Past to the Future of the Mediterranean." In *The Mediterranean: Environment and Society,* eds. Russell King, Lindsay Proudfoot, and Bernard Smith, 300–305. London: Arnold.
>   A general discussion and evaluation of Mediterranean identity, resource shortage, environmental stress, and the contradictions of development.

Pyne, Stephen J. 1997. "Eternal Flame: Fire in Mediterranean Europe." In *Vestal Fire: An Environmental History, Told through Fire, of Europe and Europe's Encounter with the World,* 81–146. Seattle: University of Washington Press.
>   Pyne is the authoritative environmental historian of fire and its effects. This chapter forms part of a volume, one in a series covering many parts of the world, and analyzes the role of fire in a very fire-prone region, namely Mediterranean Europe.

Quataert, Donald. 2000. *The Ottoman Empire, 1700–1922.* Cambridge: Cambridge University Press.
>   This book by a respected scholar strikes a balance among social, economic, and political histories, examining the major trends during the latter years of the empire; it pays attention to gender issues and to hotly debated topics such as the treatment of minorities.

Ramage, Edwin S. 1983. "Urban Problems in Ancient Rome." In *Aspects of Graeco-Roman Urbanism: Essays on the Classical City,* ed. Ronald T. Marchese, 61–92. Oxford: B.A.R.
>   This article shows that urban problems were, to a large degree, also environmental problems.

Redman, Charles L. 1999. *Human Impact on Ancient Environments.* Tucson: University of Arizona Press.
>   An archaeologist's examination of what modern society can learn from the physical record of human experience with the natural environment in the distant past. Redman emphasizes the relevance of ancient times to modern environmental problems. Countering the prevalent assumption that ecological problems are a modern phenomenon, and that ancient peoples lived in Edenic harmony with nature, he stresses that there are many cases in which early adaptations degraded the environment.

Rendell, Helen. 1997. "Earth Surface Processes in the Mediterranean." In *The Mediterranean: Environment and Society,* eds. Russell King, Lindsay Proudfoot, and Bernard Smith, 45–56. London: Arnold.

This article points out that the long history of occupation in the Mediterranean has meant that virtually all cultivated land is exploited. It investigates topics such as desertification and land degradation, erosion, the creation of badlands, and channel processes.

Rice, E. E. 1983. *The Grand Procession of Ptolemy Philadelphus.* Oxford: Oxford University Press.

> Ruler of Egypt in the third century BC, Ptolemy II held a great parade with large numbers of beasts, including many from Africa and a number from India, indicating the importation of exotic animals in the Hellenistic period.

Ringrose, David R. 2001. *Expansion and Global Interaction, 1200–1700.* New York: Longman.

> This book shows how a series of autonomous societies became interdependent on a global scale by 1700. Examining five major areas of conflict ranging from Imperial China to the Aztec and Inca Empires, it illustrates how political, cultural, and economic zones of influence expanded and overlapped. By 1700, Europeans were influential across the globe, but dominant only in a few areas, and their influence in the nineteenth century would have been hard to predict.

Roaf, Michael. 1990. *Cultural Atlas of Mesopotamia and the Ancient Near East.* New York: Facts on File.

> This atlas covers the historical development of the ancient Near East from Palaeolithic times to Alexander the Great and is illustrated with color photos of artifacts, stone inscriptions, and maps. At appropriate points in the book, special features are included that focus on specific archaeological sites around the Middle East as well as on significant cultural and scientific developments that occurred.

Rosenblum, Mort. 1996. *Olives: The Life and Lore of a Noble Fruit.* New York: North Point Press.

> Describes the habits, culture, and uses of one of the "Mediterranean triad," the three basic agricultural staple food sources of the Mediterranean lands.

Rostovtzeff, Mikhail. 1957. *The Social and Economic History of the Roman Empire,* 2nd ed., 2 vols. Oxford: Clarendon Press.

> A standard history of the economic and social development of the Roman Empire from the Julio-Claudians to Diocletian.

Sadiq, Muhammad, and John C. McCain. 1993. *The Gulf War Aftermath: An Environmental Tragedy.* Dordrecht: Kluwer Academic.

A definitive study of the environmental damage caused by the first Gulf War (1991).

Saggs, H. W. F. 1989. *Civilization before Greece and Rome.* New Haven: Yale University Press.

This study includes the predominant cultures of the ancient Near East: Babylonian, Assyrian, Egyptian, Palestinian, Sumerian, Persian, Hittite, Hurrian, and Indus Valley.

Sallares, Robert. 1991. *The Ecology of the Ancient Greek World.* Ithaca: Cornell University Press.

Sallares asks for, and provides, a rigorously ecological view of Greek history that calls into question many ideas drawn from Classical literature that have dominated historical thought. He favors biological causation and is suspicious of the idea that technology, or any form of conscious human control, guides events in the long run.

———. 2002. *Malaria and Rome: A History of Malaria in Ancient Italy.* Oxford: Oxford University Press.

A study of the evidence for the prevalence of one of the most important diseases with an ecological vector in the ancient Roman world.

Sandars, N. K., ed. 1972. *The Epic of Gilgamesh: An English Version with an Introduction.* London: Penguin Books.

One of the best available modern translations/paraphrases of the ancient Sumerian epic, which contains abundant material for interpretation of early attitudes toward the natural environment.

Santmire, H. Paul. 1985. *The Travail of Nature: The Ambiguous Ecological Promise of Christian Theology.* Philadelphia: Fortress Press.

A sweeping investigation of the Christian theology of nature, from the New Testament through the early church fathers, Augustine, Aquinas, Francis, the Reformation, and twentieth-century Protestant and Catholic thinkers.

Saunders, Corinne J. 1993. *The Forest of Medieval Romance: Avernus, Broceliande, Arden.* Cambridge: D. S. Brewer.

Western environmental thought was deeply influenced by the image of the forest in the medieval romantic writings explicated in this interesting study.

Saunders, J. J. 1965. *A History of Medieval Islam.* New York: Barnes and Noble.
   An introduction to the history of the Muslim East from the rise of Islam to the Mongol conquests. It explains and indicates the main trends of Islamic historical evolution during the Middle Ages.

Scarborough, John. 1984. "The Myth of Lead Poisoning among the Romans: An Essay Review." *Journal of the History of Medicine and Allied Sciences* 39: 469–475.
   The historiography of lead poisoning, one of the causative factors often advanced to explain the decline and fall of the Roman Empire.

Schevill, Ferdinand. 1961. *History of Florence from the Founding of the City through the Renaissance.* New York: Frederick Ungar.
   This comprehensive early history of Florence includes cultural, political, and economic developments.

Schiavone, Aldo. 2002. *The End of the Past: Ancient Rome and the Modern West.* Cambridge, MA: Harvard University Press.
   A very useful recent study of the influence of Rome on the modern world, offering a stimulating opportunity to view modern society in light of the experience of antiquity.

Semple, Ellen Churchill. 1931. *The Geography of the Mediterranean Region: Its Relation to Ancient History.* New York: Henry Holt and Company.
   A dated but still useful survey of Mediterranean geography, including human relationships. The four parts of this book are general geographic conditions, barrier boundaries of the Mediterranean region, vegetation and agriculture, and maritime activities.

Simkhovitch, Vladimir Grigorievitch. 1921. "Rome's Fall Reconsidered." In *Toward the Understanding of Jesus and Other Historical Studies,* 84–139. New York: Macmillan.
   Simkhovitch argues that the fall of Rome was caused by a failure of the agricultural base of the economy due to the unavailability of investment.

Simmons, Ian G. 1993. *Environmental History: A Concise Introduction.* Oxford: Blackwell.
   A succinct, scientifically based introduction to the field of environmental history.

Simpson, James. 1995. *Spanish Agriculture: The Long Siesta, 1765–1965.* Cambridge: Cambridge University Press.

Simpson argues for the persistence of traditional agriculture in Spain in the face of the growth of the population and the market economy, with modernization occurring only in the early twentieth century.

Smith, Catherine Delano. 1979. *Western Mediterranean Europe: A Historical Geography of Italy, Spain, and Southern France since the Neolithic.* London: Academic Press.

By drawing widely on historical and geographical sources, the author provides a synthesis of the environmental history of Italy, southern France, and Spain. She begins with people and proceeds to land use and environmental changes.

Soffer, Arnon. 1999. *Rivers of Fire: The Conflict over Water in the Middle East.* Lanham, MD: Rowman and Littlefield.

The nations of the Middle East, facing problems of population growth and the production of food and energy, have proceeded with the development of water resources without considering their neighbors' needs. This book looks at controversies over the Nile, Tigris-Euphrates, Jordan, and Orontes rivers.

Sorrell, Roger D. 1988. *St. Francis of Assisi and Nature: Tradition and Innovation in Western Christian Attitudes toward the Environment.* New York: Oxford University Press.

Sorrell, in a beautifully explained study, places Francis firmly in the context of his own times but also delineates the originality of this extraordinary figure. Francis's contributions lie in realizing nature mysticism, familial relationships with other creatures, and the performance of good works to benefit wild and tame animals.

Squatitri, Paolo. 1998. *Water and Society in Early Medieval Italy.* Cambridge: Cambridge University Press.

This well-written, archaeologically based study demonstrates that with the destruction of Roman aqueducts, early medieval Italian communities improvised locally based water supplies.

Swearingen, Will D., and Abdellatif Bencherifa, eds. 1996. *The North African Environment at Risk.* Boulder, CO: Westview Press.

A collection of authoritative articles on environmental problems in modern

North Africa, emphasizing desertification and pollution. Causative factors such as unsustainable development, population pressure, resource depletion, urbanization, and poverty are discussed.

Sweeney, Del, ed. 1995. *Agriculture in the Middle Ages.* Philadelphia: University of Pennsylvania Press.
Papers from two conferences covering agricultural development and rural society from the Carolingian renaissance to the end of the Middle Ages, including artistic and literary information.

Tal, Alon. 2002. *Pollution in a Promised Land: An Environmental History of Israel.* Berkeley and Los Angeles: University of California Press.
This environmental history of one of the most controversial Mediterranean nations traces the development that made the desert blossom as the rose, but also created air and water pollution and habitat destruction. The author has researched the history of environmentalism in Israel and speculates as to whether the present ecological disarray of the land can be redeemed.

Thirgood, J. V. 1981. *Man and the Mediterranean Forest: A History of Resource Depletion.* London: Academic Press.
A brief but dependable forest history of the Mediterranean world. Thirgood gives particular attention to Cyprus, where he had considerable field experience and responsibility.

Thomas, W. L., Jr., ed. 1956. *Man's Role in Changing the Face of the Earth.* Chicago: University of Chicago Press.
A groundbreaking series of articles from a pioneering conference on historical and contemporary environmental issues, which was consciously inspired in retrospect by George Perkins Marsh's questions as to whether humankind is an unprecedented disturber of nature's harmonies. In the 1950s, many of the ideas discussed in this collection were new and even startling. Several articles deal with Mediterranean history.

Tringham, Ruth. 1971. *Hunters, Fishers, and Farmers of Eastern Europe, 6000–3000 BC.* London: Hutchinson and Company.
An archaeologically based study of preurban history that includes a treatment of the Balkans, particularly the Danube Valley, where Tringham finds that an early bronze industry collapsed due to deforestation and consequent fuel shortage.

Turner, B. L., William C. Clark, Robert W. Kates, John F. Richards, Jessica T. Mathews, and William B. Meyer, eds. 1990. *The Earth as Transformed by Human Action: Global and Regional Changes in the Biosphere over the Past 300 Years.* Cambridge: Cambridge University Press.

> A systematic worldwide collection of studies of environmental changes. This is an intended sequel to the 1956 Thomas edition listed above but is more comprehensively organized and is mostly limited to the eighteenth, nineteenth, and twentieth centuries.

United Nations Educational, Scientific, and Cultural Organization (UNESCO). 1977. *Mediterranean Forests and Maquis: Ecology, Conservation, and Management.* Paris: UNESCO.

> Studies of conservation and ecology in Mediterranean and Mediterranean-like forest habitats around the world, including parts of South Africa, Chile, California, Western Australia, and the Mediterranean proper.

United Nations Environment Programme (UNEP). 1992. *World Atlas of Desertification.* London: Edward Arnold.

> Places the Mediterranean within the context of various forms of desertification occurring around the globe.

————. 2001. *The Mesopotamian Marshlands: Demise of an Ecosystem.* Geneva: UNEP.

> A clear view of the deliberate destruction of the Marsh Arab communities and the wetland ecosystem that was their environment and homeland.

————. 2003. *Desk Study on the Environment of Iraq.* Nairobi: UNEP.

> An objective study, well illustrated with photographs including aerial and satellite images, maps, and tables, and covering some of the environmental impacts of the second Gulf War in 2003.

van de Mieroop, Marc. 1999. *The Ancient Mesopotamian City.* London: Oxford University Press.

> A valuable introduction to the origins of urban history in Mesopotamia, including social, political, economic, intellectual, and artistic developments.

Varela-Ortega, Consuelo, José Sumpsi, and María Blanco. 2002. "Water Availability in the Mediterranean Region." In *Nature and Agriculture in the European Union: New Perspectives on Policies That Shape the European*

Countryside, eds. Floor Brouwer and Jan van der Straaten, 117–140. Cheltenham, UK: Edward Elgar.

> Water is a limited and limiting resource in the Mediterranean, and its distribution and use are surveyed in this article.

West, Louis C. 1951. "The Coinage of Diocletian and the Edict on Prices." In *Studies in Roman Economic and Social History,* ed. P. R. Coleman-Norton, 290–302. Princeton: Princeton University Press.

> The Edict on Prices is the most important extant economic document from the period of Diocletian's reforms in the late Roman Empire. This article relates it to numismatic material from the same period.

White, Gilbert F. 1988. "The Environmental Effects of the High Dam at Aswan." *Environment* 30 (September): 4–40.

> Written by a consultant to the project, this is an evenhanded analysis of the positive and negative results of the Aswan High Dam on the natural environment and people of Egypt.

White, Kenneth D. 1970. *Roman Farming.* Ithaca: Cornell University Press.

> The classic standard study of Roman agricultural methods and equipment.

———. 1984. *Greek and Roman Technology.* Ithaca: Cornell University Press.

> A useful survey of ancient technology in the Mediterranean area and Europe.

White, Lynn. 1962. *Medieval Technology and Social Change.* Oxford: Oxford University Press.

> This influential book analyzes the widespread effects of technological changes, such as the improvements of water mills and reapers, on the development of feudalism and the manor system. White also advances the controversial "stirrup theory" of the revolutionization of warfare and of all medieval society.

———. 1967. "The Historical Roots of Our Ecologic Crisis." *Science* 155: 1203–1207.

> One of the earliest and most widely known articles on the influence of Christian ideas on the development of science, technology, and environmental destruction in the medieval and modern West. White's thesis is that Christianity desacralized nature and encouraged its manipulation and exploitation by humans.

Wickham, C. J. 1988. *The Mountains and the City: The Tuscan Apennines in the Early Middle Ages.* Oxford: Clarendon Press.
> An anthropological study of a local society (two mountain valleys in central Italy) and its environment, viewed from the standpoint of ordinary people rather than the rulers.

Willcocks, William. 1903. *The Nile Reservoir Dam at Assuan and After.* London: E. & F. N. Spon.
> A description of the first Aswan Dam at the time of its construction by the engineer responsible.

Williams, Martin A. J., and Robert C. Balling Jr. 1996. *Interactions of Desertification and Climate.* London: Arnold.
> A survey of desertification in semiarid and arid regions, including the Mediterranean lands, with recommendations for management of dryland regions and indications of directions for future research.

Williams, Michael. 2003. *Deforesting the Earth: From Prehistory to Global Crisis.* Chicago: University of Chicago Press.
> A global forest history, inclusive in both chronological and geographical dimensions, but nonetheless rich in detail and attention to sources.

Winiwater, Verena. 2000. "Soils in Ancient Roman Agriculture: Analytical Approaches to Invisible Properties." In *Shifting Boundaries of the Real: Making the Invisible Visible,* eds. H. Novotny and M. Weiss, 137–156. Zürich: Hochschulverlag.
> A guide to soil nomenclature and soil testing systems in ancient Latin sources.

Wink, Andre. 2002. "From the Mediterranean to the Indian Ocean: Medieval History in Geographic Perspective." *Comparative Studies in Society and History* 44, 3: 416–445.
> A study of trade and intercultural contact across the Middle East in the Middle Ages.

Young, Gavin. 1977. *Return to the Marshes: Life with the Marsh Arabs of Iraq.* London: William Collins Sons.
> A description of Marsh Arab environment and culture in the days before the lamentable destruction of land and society.

Zaidi, Iqtidar H. 1986. "On the Ethics of Man's Interaction with the Environment: An Islamic Approach." In *Religion and Environmental Crisis*, ed. Eugene C. Hargrove, 107–126. Athens: University of Georgia Press.

This study indicates that Islamic tradition assigns the responsibility for proper environmental protection and regulation to the Islamic state.

Zupko, Ronald Edward, and Robert Anthony Laures. 1996. *Straws in the Wind: Medieval Urban Law in Northern Italy.* Boulder, CO: Westview Press.

The authors show that Italian city-states were aware of the problems of urban pollution and took steps to prevent it by legal enactments and enforcement.

INDEX

ABOUT THE AUTHOR

J. Donald Hughes, Ph.D., is John Evans Professor of History at the University of Denver in Denver, Colorado, and is one of the founders of both the American Society for Environmental History (ASEH) and the European Society for Environmental History. He is a past editor of the journal *Environmental Review* (now *Environmental History*). His published works include *Pan's Travail: Environmental Problems of the Ancient Greeks and Romans* (1994) and *An Environmental History of the World: Humankind's Changing Role in the Community of Life* (2001). He received the Distinguished Service Award of the ASEH in 2000.